Praise for Zero to GenAI Prod

"AI is transforming how product managers work and what the role means. Product managers who want to thrive in this new era should read Saumil Shrivastava's comprehensive guide full of useful advice." — **Dan Olsen**, author of *The Lean Product Playbook*

"Zero to GenAI Product Leader offers a clear, practical look at how generative AI is reshaping product management. It connects the dots between emerging AI capabilities and what they mean for product decisions, making it a useful resource for PMs who want to stay current and think more strategically about this fast-moving space." — **Lewis C. Lin**, *Author of Decode and Conquer*

"In a space crowded with hype, this book stands out for its clarity and depth. It's a practical, strategic roadmap — whether you're transitioning into GenAI product management or scaling impact in your current role. If you want to develop the mindset, systems thinking, and product discipline needed to succeed in this new era, start here." — **Matt Schnugg**, *Chief Product Officer, Digital Grid @ Schneider Electric (formerly at Google Cloud, GE Digital, Microsoft)*

"*Zero to GenAI Product Leader* is an essential guide for anyone building AI-powered products, blending rich historical context with practical frameworks tailored to today's fast-evolving landscape. It demystifies both generative AI and product management, then bridges them with clear insights into emerging business models, real-world applications, and the unique skills PMs need to succeed in this space. With compelling case studies and actionable advice, this book is a must-read for product leaders, founders, and teams shaping the future of AI-driven innovation." — **Steve Sweetman**, *Partner PM, Azure OpenAI*

"This is a timely and much-needed read for product managers navigating the GenAI era. The pace of change is fast, and success demands more than technical fluency. It requires grounded product thinking, clear decision frameworks, and the ability to build responsibly at scale." — **Vamsi Krishna**, *Co-founder & CEO, Vedantu — one of India's leading live online learning platforms*

"Want to AI-proof your career? Become an AI maker, not just an AI consumer. Saumil Shrivastava shows you exactly how to do that, because he's made the journey himself." — **Jeremy Schifeling**, *Author of* Career Coach GPT, *the #1 AI-for-careers best-seller.*

"As GenAI evolves into Agentic AI, product managers must go beyond building tools — they must orchestrate intelligence. That shift requires not just technical intuition, but product discipline, cross-functional leadership, and an eye on long-term business impact. Zero to GenAI Product Leader is a rare guide that doesn't just follow the wave — it prepares you to lead it." — **Suvrat Joshi**, *Senior Vice President, Product Management at Nintex (formerly CPO at Veriff, Product Leader at Dropbox, Meta and Yahoo)*

"From startups to Big Tech, GenAI is rewriting what it means to be a product leader. You don't need an AI degree to get started — but you do need curiosity, clarity, and a strong roadmap. This book delivers exactly that: practical guidance for breaking into GenAI product management, strategic depth for leveling up, and the perspective PMs need to lead with confidence in a fast-moving, ambiguous landscape." — **Amit Ghorawat**, *Director of Product Management @ Reddit (formerly at Google, BCG, and Harvard Business School alum)*

"This book is more than just a guide to GenAI — it's a lens into the future of product leadership. As AI becomes increasingly embedded in consumer and enterprise experiences, PMs need to evolve from feature owners to system orchestrators. Saumil's work is a thoughtful, well-structured playbook for making that shift with clarity and confidence." **— Abhishek Sharma**, *Product Leader at Meta (formerly Head of Product at Amazon and Senior Product Manager at Samsung)*

"The landscape of generative AI is not just new; it is shifting beneath our feet at incredible speed. This book is the essential field guide for the product pcapter 6ioneers of this volatile new era. It offers the grounding principles needed to navigate the rapids and chart a course toward durable, human-centric innovation amidst the accelerating change" — **Sunny Tahilramani,** *Product Management Leader, Generative AI at Google (Formerly at UiPath, Microsoft)*

"From building personalized home design tools to developing reasoning-capable AI agents, I've seen firsthand how GenAI is evolving from content generation to intelligent action. This book doesn't just prepare PMs for that shift — it equips them to lead it. A must-read for anyone serious about agentic AI and the next frontier of product innovation." — **Partho Sarkar**, Agentic AI Builder at Nurix (formerly Co-founder & CTO at Styldod)

"*Zero to GenAI Product Leader* isn't just a roadmap to mastering emerging technology — it's a masterclass in balancing bold innovation with the disciplined rigor required to earn user trust. If you're looking to lead with both vision and responsibility, this book is your essential guide." — **Selena Zhang**, *AI Product Leader at a leading global social media company (formerly at Amazon)*

"*Zero to GenAI Product Leader* demystifies what it really takes to become a GenAI Product Manager. It doesn't just teach the tech — it teaches the thinking. If you're serious about building or advancing your AI product career, this book is your playbook. Strongly recommended for aspiring and experienced PMs alike." — **Rocky Zhang**, *AI Product Leader at Airbnb (formerly at Redfin and Amazon)*

Zero to GenAI Product Leader

The Complete playbook for AI product management in the GenAI and Agentic AI era

———

Saumil Shrivastava
Published by Abir LLC
Kirkland, WA
2025

Note to the Reader

Writing this book has been one of the most meaningful, challenging, and humbling experiences of my career. But I didn't write it to share everything I know. I wrote it because this field is evolving too fast for any one of us to have it all figured out. We're all learning in real time. That's what makes this work so energizing—and so human.

Throughout this journey, I collaborated with AI tools, including agents, to accelerate research, spark new ideas, and refine my thinking. In many ways, this book is a small example of what becomes possible when we build *with* AI, not just *for* AI.

If there's one hope I have for you, it's this: that you walk away not just feeling more prepared, but more empowered. More confident in your voice. More willing to take risks, ask better questions, lead with curiosity, and stay grounded in what matters most.

AI will transform industries. But it's product managers who will play a key role in deciding how that transformation shows up in people's lives—what gets built, who it's for, and what it leaves behind.

That's not a small responsibility. But you're not alone in carrying it. You're part of a community of builders—people who believe that product is more than velocity or metrics. It's about vision. Craft. Care.

Thank you for giving this book your time and attention. I hope it becomes a reference point, a confidence boost, or a spark when you need it most.

And wherever this journey takes you next—I'll be cheering you on.

Let's keep building.

Note: The views and opinions expressed in this book are solely my own and do not reflect those of my employer.

Dedication

To *Mili*,

my partner, my greatest supporter, and the quiet force behind this book.

When I doubted myself, you reminded me why this work mattered. When I faltered, you held up the mirror to show me what I was capable of. You didn't just encourage me to keep writing, you took on the weight of bringing this book to the world, building the bridge between my words and the people who would read them. Every page of this book carries your fingerprints in ways most will never see, but I will never forget.

You made it possible, and you made it better.

And to *Abir,*

born in 2018, at the very beginning of this long journey. You grew alongside this book, and every day you reminded me what it means to dream big, to stay curious, and to keep going, no matter how hard it gets. You are my inspiration for what it means to build with love, with care, and with a vision for something better.

Both of you are the reason this book exists.

And to **all the dreamers, builders, and future GenAI Product Leaders out there —**

those who dare to walk the lonely path of creating something that didn't exist before,
those who believe in the power of imagination even when no one else does,
those who carry the weight of doubt and still choose to move forward
this book is for you, too.

Keep building. The world needs what you dream.

Preface

This book began with a simple idea: to help people **break into product management** with clarity, confidence, and structure, especially in a world now transformed by **Generative AI**.

Whether you're an aspiring product manager, an experienced PM wanting to scale your impact in the AI era, a career switcher from engineering, design, data science, or a non-technical background, or a founder building AI-native products, this book is designed to give you the mental models, tools, and confidence to thrive. It's for people starting from scratch and for PMs already in the role who want to lead in AI-driven environments.

Over the years, I've coached aspiring product managers on everything from résumé reviews and mock interviews to navigating career transitions into tech. During my time at the Ross School of Business, I led a cohort of first-year students through the same journey—helping them land their first product roles in technology companies. Through these experiences, I saw the same questions come up again and again: *How do I break in? How do I grow once I'm in the role? How do I stay relevant when the ground keeps shifting?*

While there was plenty of advice floating around, there wasn't a single, structured, actionable guide that brought it all together. Most people were piecing together YouTube videos, scattered blog posts, and half-answered Reddit threads—unsure where to start or how to make meaningful progress.

I wanted to change that. So in late 2017, I began writing. Then life intervened. The pandemic brought unpredictable shifts - remote work, health scares, and the unexpected joy of more time with my young son. Writing took a back seat. The book remained unfinished.

In 2022, I returned to the manuscript with renewed focus. But the world had already shifted again.

Generative AI had gone mainstream. Suddenly, everyone was talking about it - at work, at home, in headlines, and in hallway conversations. People were eager to understand **GenAI** and **Agentic AI**: how to use them, how companies were adopting them, and most of all, how they would impact their careers.

I experienced that shift firsthand. After leaving my previous role amidst broader industry changes, I entered a job market where roles I once felt qualified for now demanded fluency in AI technologies. With widespread layoffs and heightened competition, **breaking into AI product management** felt overwhelming.

At the time, I had little direct experience in AI. I wasn't sure where to begin. I faced rejections, struggled to upskill, and fumbled through a maze of fragmented resources. Ironically, I began using AI itself to learn faster. After months of effort and uncertainty, I landed a role on the **AI Platform** team at Microsoft—my dream job, and one that placed me at the heart of AI innovation.

And once again, people began reaching out, just like they had in the early days:
"How can I become a GenAI PM?"
"Where do I start?"
"Will I become irrelevant if I don't learn AI?"

But just like before, I didn't have a single, easy answer.

That's when I knew the book had to evolve. It could no longer focus solely on breaking into traditional product management. It had to reflect the new reality, where product managers must blend timeless product skills with the accelerating demands of **Generative AI** and **Agentic AI**. This isn't just for those trying to land their first AI product role. If you're already a PM, this book will help you sharpen your product sense, master AI-era decision-making, and lead with confidence as the role evolves.

This book isn't a brain dump of everything I know. It's a synthesis of what I've learned—through coaching, failure, experimentation, and career transition. It weaves together insights from mentoring others and navigating my own journey into **AI product management** in a rapidly shifting landscape.

It's written for anyone who wants to thrive in the AI era of product leadership—whether you're an aspiring product manager looking to break into **AI-powered product roles**, an experienced PM determined to stay relevant, a founder building **AI-native products**, or a career switcher coming from engineering, design, data science, or even a non-technical background. If you're curious about moving into **AI-driven product roles**, this book bridges the gap—helping you understand how AI is reshaping the PM role, and giving you the tools, frameworks, and mental models to make the transition with confidence.

Some parts—like building product sense, structuring interviews, and leading teams—will remain relevant for years. Others will evolve as AI transforms how we build, measure, and deliver products. That's why I've structured the book in two parts. **Part I** covers the theory: the foundations of Generative AI, Agentic AI, the GenAI product lifecycle, emerging AI business models, and the skills today's AI product leaders need. **Part II** is about practice: decision-making frameworks, career transitions, interview playbooks, and staying ahead in a market moving at breakneck speed.

Rather than trying to capture a frozen snapshot, I've built a durable foundation— one that adapts with you. That's why I consider this a beta version, not because it's unfinished, but because adaptability is the only way to stay ahead in AI. This book is a starting point, not an endpoint.

To help readers stay updated beyond the book, I'll be launching a Substack newsletter alongside its release. It will feature timely insights, curated **GenAI PM interview tips**, and perspectives from professionals navigating the shift into **AI product roles**. As a reader, you'll receive complimentary access to the free tier. For details on how to subscribe, see Appendix F: *How to Stay Updated.*

This isn't just a guide - it's the **AI product management playbook** I wish I had when I was making my own career switch into the AI era. Whether you're starting fresh, transitioning from another role, or trying to future-proof your career, I hope it demystifies the journey and gives you clarity, confidence, and momentum to move forward.

Table of Contents

Part I · Theory

Part II · Practice

Chapter 1: Introduction

Maybe you've been hearing a lot about Generative artificial intelligence (GenAI) lately—how it's changing the way we work, think, create. And maybe somewhere in that buzz, the idea of becoming a GenAI product manager caught your attention.

But now you're wondering: Where do I even start?

You might come from engineering, design, marketing, or business—perhaps with years of experience in the tech industry, or maybe from a completely different world altogether. Or you might already be working in product management, starting to feel the ground shift beneath your feet. Or maybe you've never held the title of product manager, but you've always found yourself thinking about how products should work, how to make them better, and how emerging technologies could shape what comes next.

I've had dozens of conversations that started just like this with friends, students, colleagues, people asking the same questions: *"Could I become a PM in the age of GenAI?*

The answer? Yes, you absolutely can.

You don't need to be an artificial intelligence researcher or machine learning expert to step into GenAI product management. What you *do* need is curiosity, adaptability, and a strong foundation in product thinking—things like understanding users, solving real problems, and working collaboratively to deliver meaningful outcomes. These timeless skills remain at the heart of great product management, even as the landscape evolves.

What's changing is the context. GenAI is rewriting how products are conceived, built, and experienced at a pace that's hard to ignore. Companies aren't just seeking technical expertise; they're looking for PMs who can bridge the gap between cutting-edge technologies and real human needs. Those who can translate the potential of artificial intelligence (AI) into practical, responsible solutions are becoming increasingly essential.

This shift is opening new doors for professionals across backgrounds, but it also demands a reset. Whether you're starting fresh or shifting from an adjacent role, your first step into this new world is the same: grounding yourself in what product management truly involves. That's where this chapter comes in.

We'll break down the core responsibilities of a PM, explore the different types of roles that exist, and outline the skills that matter most. Whether you're exploring product management for the first time or brushing up before making the leap into GenAI, this foundation will serve you well.

Because once you understand the role clearly, you'll be ready to explore how GenAI is reshaping the role and how you can lead in this new era. Let's begin.

The World of Product Management

Are you someone who thrives on solving problems, someone who can't help but imagine better ways to do things? Do you find energy in collaboration, in bringing different minds together to create something that didn't exist before? If the idea of shaping products from a simple spark of inspiration to something people actually use and love excites you, then product management might be a great career for you.

Product management isn't just a job title, it's a mindset. It's about seeing opportunities where others see complexity and bringing the right ideas to life. Whether you're an engineer who wants to move beyond coding, a marketer drawn to building smarter experiences, or someone from a non-tech background looking for an entry point, understanding the essence of this role will give you a solid foundation as you prepare to navigate the GenAI era.

So, let's start with the basics, what is product management?

What Is Product Management?

Think about the products you use every day like a chatbot that helps resolve a customer support issue, or a recommendation engine that somehow knows what you want to watch next.

Behind every great product is a PM helping guide its development, figuring out what to build, why it matters, and how to deliver value for both the user and the business.

At its core, product management is about solving meaningful problems. It's spotting what users need, sometimes before they can say it, and working with a team to bring that solution to life. PMs don't always write code or design interfaces, but they help steer the ship: asking hard questions, making trade-offs, and aligning everyone toward a shared outcome.

In real-world products, their impact can be subtle but essential. At Netflix, for instance, a PM might focus on how the "watch next" queue adapts to viewing habits, working with engineers and data scientists to improve discoverability without overwhelming the user. At Zoom, another PM might be helping reduce audio lag, digging into the root cause and working across teams to prioritize improvements that affect live calls. In both cases, the PM's job is to make sure the right problem is being solved in the right way. That means clarifying intent, removing ambiguity, and helping the team stay aligned as they move from idea to execution.

And sometimes, product management is about rethinking the experience altogether. Peloton didn't just release a bike, they created a system that made solo workouts feel social, structured, and habit-forming. Features like live classes, progress tracking, and instructor-led engagement weren't just technical implementations, they were product decisions grounded in user behavior, motivation, and long-term stickiness.

What makes a PM effective isn't just their ability to ship, it's their ability to shape. Notion changed how people structure knowledge across teams. Duolingo made language learning part of daily routine through gamification. These aren't just feature sets, they're shifts in behavior. PMs helped drive those shifts by listening closely, making trade-offs visible, and aligning design, engineering, and business toward a shared outcome. That's where the craft gets real: when you can turn a fuzzy problem into something clear, buildable, and valuable.

To do that well, great PMs consistently balance three forces—desirability (do users want it?), feasibility (can we build it?), and viability (will it work for the business?). It sounds simple, but holding those forces in tension is what makes the job so challenging—and rewarding.

As Josh Elman, a seasoned product manager who helped shape platforms like Twitter, LinkedIn, and Facebook, puts it, the job of a product manager is to

> *"Help your team (and company) ship the right product to your users."*

That captures the heart of the job—but let's break it down. *"Help your team"* isn't about directing from above, it's about enabling progress. As Josh Elman describes it, product managers drive alignment, remove blockers, and keep the team focused on what matters most.

"(And company)" is a reminder that your team's work should serve the company's broader goals. You're not building in isolation, you're connecting day-to-day execution to long-term strategy.

"Ship" signals momentum. Shipping doesn't mean perfect, it means delivering value, learning quickly, and improving over time. PMs balance that urgency with thoughtfulness.

"The right product" means something that solves a real user problem and makes sense for the business. It's not about building more, it's about building what's meaningful.

"To your users" centers the work where it belongs: with the people you're solving for. PMs are advocates for users throughout the process, translating their needs into real impact.

You're not just managing timelines, you're guiding a product from idea to outcome, from ambiguity to alignment. And when done well, that impact scales beyond the roadmap, it shapes how teams build, how users live, and how companies grow.

And now, that job is evolving. GenAI is reshaping product management—not replacing its foundations but expanding what's expected. The mindset we just explored still matters: solving the right problems, aligning teams, and shaping user behavior. But what does that actually look like in practice? Whether you're new to the field or shifting into GenAI roles, it helps to understand the building blocks of the job. Because in a space that's moving this fast, clarity on your role becomes even more important.

So let's break it down. What responsibilities do PMs typically own and how do those responsibilities shift depending on the product, the company, or the moment in the product's life cycle?

Key Responsibilities of a Product Manager

Most product managers take on a common set of responsibilities but how those responsibilities play out can vary based on several factors: the type of product you're building, the size of the company, the industry you're in, the stage of the product, or even the culture of the organization.

In smaller teams or startups, a single PM might wear many hats, owning everything from user research to delivery and go-to-market planning. The pace tends to be faster, decisions more informal, and roles more fluid. In contrast, larger organizations often distribute these responsibilities across specialized teams. You may work within more structure, but your role in connecting strategy to execution—and keeping the product vision coherent—becomes even more critical.

While company size influences how broadly a PM's responsibilities are distributed, the type of product or industry can shape what you focus on most. A PM working on infrastructure might prioritize reliability and developer APIs. A consumer PM may zero in on engagement, usability, and fast iteration. And in highly regulated spaces like finance or healthcare, PMs often spend more time aligning with compliance and legal teams. These shifts change how the job feels day to day—but the core responsibilities remain the same.

And in today's GenAI landscape, where even established companies are moving with startup speed, the ability to adapt quickly, go hands-on, and operate across disciplines is more valuable than ever. You might find yourself writing scripts to test AI system behavior, working with AI researchers, or rapidly prototyping new interactions. We'll explore these emerging skills more in Chapter 4.

For now, let's look at the foundation: the core responsibilities PMs take on across different products, companies, industries, and team setups.

Understanding the market

Strong product decisions begin with a clear understanding of the market and that doesn't just mean knowing what your competitors are doing. It means staying close to your users, tracking their behaviors and unmet needs, and spotting the broader shifts that signal where things are heading.

This work isn't a one-time research sprint, it's ongoing. You're constantly talking to users, reviewing feedback, analyzing trends, and looking for patterns in how people use (or struggle to use) your product. Often, what matters most isn't what users say, it's what they do, and where they hesitate, switch tools, or give up entirely.

For example, when the team behind Notion noticed users toggling between multiple apps—one for notes, another for tasks, a third for spreadsheets—they didn't just optimize their note-taking features. They stepped back and asked a more strategic question: *Could we bring all of this together?* That insight shifted the product's direction—from a note-taking tool to a flexible workspace that could replace entire workflows.

This kind of market awareness isn't just about reacting to what exists. It's about spotting opportunities in friction and using those signals to inform your roadmap. In the GenAI era, where new tools, models, and interaction patterns are emerging at a rapid pace, staying close to the market is even more essential. You're not just watching what competitors release; you're tracking open-source innovations, shifts in user expectations, and entirely new workflows made possible by AI. The landscape is moving fast, and product intuition alone isn't enough, you need ongoing signals from the real world to guide your roadmap.

Whether you're working on an internal tool or a public-facing product, understanding the market helps you ground your team in real-world needs and avoid building in a vacuum.

Defining the vision

Once you understand the market, your next job is to define where the product is going and why that direction matters. This is your product vision. It doesn't need to be flashy or complex. It just needs to give the team a clear sense of purpose. Good visions answer simple but powerful questions:

- What are we trying to solve?
- Who is this for?
- What will be different once we've solved it?

That clarity becomes your north star. It helps engineers prioritize trade-offs, designers frame the experience, and business teams connect the dots. Without a shared vision, even the most talented teams risk building in circles.

Take Figma, for example. Instead of building just another design tool, the team focused on something deeper: real-time collaboration. Designing together in the browser, just like Google Docs, became their North Star. That single shift in vision shaped everything that followed, from live cursors to developer handoff flows. It wasn't just a set of features, it was a rethinking of how design teams work together.

The same principle applies in today's GenAI products. A strong vision doesn't just describe what technology can do; it clarifies the outcome it's meant to enable. Are you helping analysts get insights faster? Reducing repetitive work for customer support? When your vision focuses on the user outcome, not just the tech, it becomes much easier to guide product decisions, especially as the underlying AI capabilities continue to evolve.

Strong product visions are specific enough to guide, but flexible enough to grow. They focus on the *why*, not the *how* because the how will often shift as you learn more. So, before you get into roadmaps or metrics, pause and ask: Do we know what we're trying to become and why it matters? That answer doesn't just align your team. It earns trust from your stakeholders, and conviction from yourself.

Managing the product lifecycle

A product goes through many phases, from early ideas to first launch, and from iterative updates to long-term growth. As a PM, your job is to manage that full journey, guiding the product through discovery, development, release, and ongoing evolution. At each stage, you're asking different questions. Early on: What problem are we solving, and for whom? During development: Are we building it in a way that's usable, performant, and scalable? After launch: How are users responding—and what needs to change to keep the product relevant and valuable?

This work doesn't stop after version 1. In fact, some of the most important decisions come later, when you start to see how the product behaves in the real world. You'll monitor adoption, analyze feedback, and stay alert to changes in user behavior, market dynamics, or tech capabilities. Sometimes that means doubling down on what's working. Other times, it means course-correcting based on what you're learning.

Spotify offers a clear example. It started as a music streaming app, but over time, the team noticed new patterns. Users wanted podcasts. They craved personalized discovery. They wanted to share experiences with friends. Instead of treating launch as the finish line, the team expanded the product's scope adding podcast support, personalized playlists like Discover Weekly, and social features like Blend. Each move reflected an ongoing response to user needs and a commitment

7

to lifecycle thinking. That's what managing the lifecycle is about: treating the product as something alive. You're not just shipping features; you're shaping how the product grows and adapts over time. And in fast-moving spaces like GenAI, where models, interfaces, and expectations shift quickly, this mindset matters even more.

Prioritization: deciding what to build next

There's never a shortage of ideas. Everyone, from users to executives to teammates, has thoughts on what the product should do next. But time, people, and resources are always limited. That's where prioritization comes in.

As a PM, your job isn't just to collect ideas, it's to make hard choices. You'll weigh impact versus effort, short-term wins versus long-term bets, and user needs versus business goals. You'll need to ask: What matters most right now? What can wait? What are we saying no to and why? This process isn't just about frameworks or scorecards (though they can help). It's about building a clear rationale. That way, when you say no or not yet, you can explain the trade-offs in a way that builds trust with your team and stakeholders.

Take Slack, for example. Should the team invest in redesigning the mobile app to improve the on-the-go experience? Or should they focus on strengthening integrations with tools like Salesforce and Jira to better serve enterprise customers? Both are worthwhile. But depending on the company's goals, user signals, and technical readiness, one will likely rise to the top. Prioritization helps you surface that clarity—and act on it.

In GenAI products, prioritization comes with new questions. Does the model produce reliable outputs for this use case? Will the experience improve as the model evolves? Is there a way to launch quickly and learn before committing deeper? These aren't just feature decisions, they're decisions about feasibility, ethics, and iteration loops. What stays the same across all products is the core skill: making thoughtful decisions with limited information, and keeping the team aligned as those decisions evolve.

Leading cross-functional teams

Product managers rarely build things on their own. Your impact comes from how well you bring people together—engineering, design, marketing, data science, legal, sales, support, and more. Regardless of the org chart, your job is to connect the dots.

This isn't about being the loudest voice in the room. It's about aligning different perspectives around a shared outcome and making sure everyone understands the "why" behind what you're building. That often means translating between worlds: helping designers understand technical constraints, or helping engineers see what really matters to the user.

It also means managing across different types of influence. You won't always have formal authority, especially with peer teams. But you will have to earn trust by being clear, prepared, responsive, and grounded in real user needs. As your product grows, so does the web of collaboration. One week, you might be working with legal to ensure data compliance. The next, you're with the marketing team preparing a launch, or coordinating across design and engineering to unblock a release.

In GenAI teams, the mix expands further. You might be collaborating with prompt engineers, AI researchers, or infrastructure leads. That adds new complexity, but it also creates new opportunities to learn, experiment, and shape entirely new types of experiences. At the heart of it, leading cross-functional teams is about clarity, context, and communication. When people understand the problem, the plan, and their role in it—they move faster, with more confidence. That's what you enable.

Types of Product Managers: How Context Shapes the Role

For a long time, product management followed a familiar structure. Your role, whether technical or non-technical, B2B or B2C, was shaped by your product's audience, your company's stage, and the lens you brought to the table. You might have focused on launching features, growing users, refining internal platforms, or crafting smooth UX flows.

These traditional paths helped define careers. They still do. They offer a solid foundation in product thinking, teamwork, and execution. Even in the GenAI era, these classifications remain useful but they're being stretched. Each one is gaining new layers as AI transforms what teams build, and how.

Generative AI is changing not just what PMs work on, but how they operate. PMs are being pulled closer to the technology itself. The line between planner and builder is getting thinner. In the chapters ahead, we'll explore how that evolution is unfolding. But first, let's look at the roles that have historically shaped the field.

That context will help you understand where you've been or where you might start before stepping into what's next.

Customer type: B2B vs. B2C

One of the biggest factors shaping a PM's focus is who the end customer is. In B2B settings, product managers often work with enterprise buyers, complex deployment needs, and multi-stakeholder decision-making. The value proposition must be clear, measurable, and often tied to business outcomes. In contrast, B2C PMs focus on the individual user experience where speed, emotion, habit formation, and usability drive adoption. The feedback loop is often faster, but so is the risk of churn.

Technical orientation: Technical vs. Non-technical

PMs also differ in how deeply they engage with the underlying technology. Technical PMs often come from engineering or data backgrounds and are comfortable diving into APIs, performance metrics, or architecture trade-offs. Their role is closely aligned with engineering. Non-technical PMs, meanwhile, may focus more on market dynamics, user insights, and business goals. They collaborate across teams but tend to lead with product sense and strategic clarity rather than technical depth.

Functional focus: Growth, Platform, Data, UX

Some product managers specialize based on the problem they're solving or the function they own. A growth-focused PM concentrates on user acquisition and retention—experimenting with flows, incentives, and behavioral nudges to drive usage. A platform PM builds tools and services that enable other teams, internal or external, to succeed. That might mean shaping developer platforms, scaling APIs, or managing core infrastructure used across products.

Data PMs focus on turning information into strategic advantage. That could involve building analytics dashboards, enabling personalization through data pipelines, or designing features that surface insights for end users. If you're working on storage systems, data services, or orchestration layers that support this kind of work, your role might lean toward platform PM, but the emphasis differs. Data PMs typically center their work on how data is used, while platform PMs focus on how data (or other systems) are delivered reliably and at scale.

UX PMs, meanwhile, are focused on interaction design and usability making sure every experience feels intuitive, efficient, and meaningful. They're the ones advocating for frictionless flows and emotional resonance in every click and scroll.

These functional paths aren't rigid—they're lenses. They reflect where your product makes the most impact, and how you, as a PM, shape that impact.

Specialized focus: Industry or domain-specific

Finally, some PM roles are defined by the environment they operate in. Working in a regulated industry like finance, healthcare, or education demands familiarity with unique constraints, user types, and compliance needs. Others focus on niche domains like security, privacy, or sustainability—where deep domain knowledge becomes essential. These roles often blend core product skills with industry fluency.

A shifting landscape

These categories aren't just job titles, they reflect real preferences, skills, and working styles. I've walked through a few of these paths myself, and each one taught me how to listen better, build smarter, and lead with more context.

But the landscape is shifting. Fast. GenAI, and increasingly Agentic AI, is rewriting the playbook. PMs are now shaping systems that can act autonomously. That might mean designing workflows where agents collaborate, tuning prompts that guide real-time behavior, or building prototypes before engineering even joins the conversation.

Still, the fundamentals haven't disappeared. Curiosity. Clarity. Collaboration. These traits remain at the core but how they show up is evolving. The classifications you just read about still matter but they no longer define the full picture. A growth PM might now be responsible for optimizing agent handoffs. A UX PM might help shape prompt responses. Every role is expanding.

Before we zoom into what's changing, let's get grounded in the traits that make strong PMs strong—across any era.

Core Competencies of a Product Manager

You've seen the paths a PM can take, finding your fit is just the start. But no matter where you land, there's a core that holds it all together, a set of strengths that make you shine. I've tripped over enough projects to know it's not about fancy titles or tools—it's about what you bring to the table. Here's what I've pieced together that makes a PM thrive, skills that stand up, even as GenAI rewrites the rules.

Strategic execution turns the big picture into reality. It's not just spotting user needs or business goals, it's cutting through the noise, picking what matters, and rallying the team to nail it. Think of refining Airbnb's booking flow, you're not just dreaming; you're making it happen, tight and focused. I've seen PMs drift without that clarity, lost in feature clutter. It's how you keep the team on track, no matter what's ahead.

Communication, with empathy, gets everyone moving together. It's listening to engineers, designers, and execs—really hearing them—while stepping into their shoes to feel their side. That's how you show why a Cerner EHR matters to doctors or pitch a growth tweak that clicks—understanding their goals, building trust. I've learned the hard way: no trust, no momentum.

Problem-solving keeps you steady when things break. Every role, growth, data, whatever hits roadblocks. It's digging into why a Duolingo streak's failing or a Slack channel's lagging—figuring the snag, fixing it, pushing on. I've watched PMs stall without that grit, it's how you turn messes into wins.

Adaptability is your resilience in a shifting world. GenAI's changing the game, think Copilot in Microsoft Teams, making hybrid work smoother overnight. New regulations like data privacy, new tech—you've got to pivot fast and stay solid. Some PMs cling to old ways and fade, but this flexibility keeps you rolling. Here are some key things to keep in mind:

- Stay curious: Dig deeper, it keeps you sharp.
- Own the trade-offs: Every yes means a no, face it clear-eyed.
- Guide, don't push: Point the way, let the team build it.

That's what I've seen carry PMs through, from my own stumbles to GenAI's wild ride. These skills are your bedrock—timeless, sure—but the PM world's bending fast. GenAI's not just another tool; it's hitting every corner of the craft, nudging us all to rethink how we build.

The GenAI Shift for Product Managers

Product management has always adapted to tech shifts. SaaS (software as a service) changed how software was delivered, cloud reshaped how we scale, and mobile redefined user expectations. But GenAI is different. It's not just another evolution, it's transforming how products are imagined, built, and improved in real time.

Until recently, AI was seen as a specialized domain, mostly relevant to AI-first companies or advanced data teams. That's no longer the case. When I was job hunting in late 2022, I noticed a shift. GenAI wasn't just a feature, it had become a core expectation in nearly every product role. It wasn't just confined to AI-driven products; it was being embedded across software, infrastructure, and digital experiences. Traditional PMs were now expected to understand GenAI and apply it meaningfully.

Traditional PM strengths, clarity, curiosity, and collaboration, still matter. But GenAI adds new demands: a need for faster iteration, deeper technical intuition, and closer collaboration with AI engineers and researchers. In many teams, the PM is no longer just shaping specs, they're helping prototype early flows, guide model behavior, and pressure-test whether something is feasible at all.

The role is becoming more hands-on. More experimental. More tightly woven into how intelligence gets built into software. But to understand why this shift is happening now, we need to take a step back. GenAI didn't appear overnight. It was built on decades of research, breakthroughs, and quiet AI advancements—long before ChatGPT made headlines.

In the next chapter we'll explore that journey, how GenAI evolved from an academic curiosity into a force that's reshaping every other industry, and the product management role with it.

Chapter 2: The Rise and Business Impact of Generative AI

Generative AI may feel like a sudden transformation, but it was decades in the making. From early rule-based AI systems to deep learning breakthroughs, each step built toward the moment AI became mainstream.

The Rise and Market Evolution of Generative AI

This section explores how generative AI emerged, gained widespread adoption, and reshaped the competitive market. We'll trace its evolution, the forces that made its breakthrough possible, and the players that accelerated its rise.

The Road to Generative AI: How We Got Here

When I first started exploring AI product management, I didn't fully grasp how deeply AI had already shaped the world around us. I knew AI-powered recommendations influenced what I watched on Netflix and what I bought on Amazon. I had seen chatbots answering customer questions and voice assistants like Alexa making everyday tasks easier. But I didn't realize that AI had been decades in the making—a story of breakthroughs, setbacks, and reinventions that had quietly transformed industries before generative AI changed everything. For many people, including myself, AI felt like something that belonged to researchers, engineers, or highly technical teams. That changed in late 2022, when ChatGPT and other generative models became part of everyday life. Suddenly, AI wasn't just making recommendations in the background, it was writing, coding, designing, and assisting in ways that felt almost human.

This shift didn't happen overnight. It was the result of a long journey, one that started over 70 years ago. Figure 2.1 illustrates a timeline of key moments in the development of artificial intelligence, beginning in 1950.

1950 – Turing Test Introduced
Alan Turing proposes a test for machine intelligence, asking: "Can machines think?"

1956 – Dartmouth Conference
AI becomes a formal academic field, sparking the first wave of AI research

1970s–1980s – AI Winter
Progress stalls due to limited computing power and unmet expectations: funding

1998–1999 – Early Emotional and Consumer AI
MIT's Kismet shows emotional intelligence; Sony's AIBO learns from user interaction

2016–Tay Controversy
Microsoft's chatbot Tay misbehaves, highlighting the importance of AI ethics

1997–Deep Blue Beats Kasparov
IBM's chess AI defeats the world champion, showcasing strategic AI capapbilities

2002–Roomba Launches
Autonomous household AI enters the mainstream through robotic vacuums

2014 – Siri and Watson Debut
Apple's Siri brings voice AI to consumers; IBM Watson wins Jeopardy! with realt·--NLP

2014–DeepMind and Alexa
Google buys DeepMind: Amazon Alexa popularizes smart home voice assistants

2020–GPT-3 Released
OpenAI's large language model stuns the world with human-like text generation

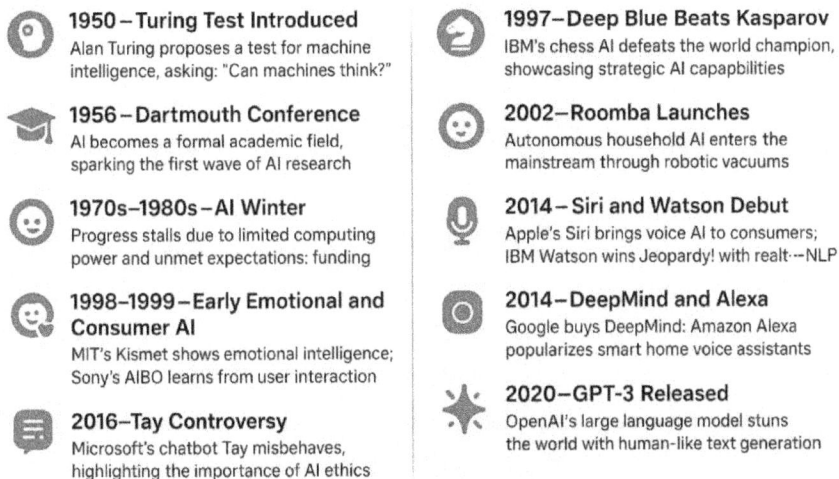

Figure 2.1. Timeline of some key moments in the development of artificial intelligence.
© 2025 by the author of *Zero to GenAI Product Leader*.

1950s–1980s: the birth of AI and the first wave of excitement

In 1950, Alan Turing posed a radical question: Can machines think? His Turing Test laid the foundation for artificial intelligence, inspiring decades of research into how machines could mimic human reasoning.

By 1956, at the Dartmouth Conference, AI became an official field of study. Researchers believed that, within a few decades, machines would match human intelligence. Early progress was promising. Expert systems were built to diagnose diseases, robots like SHAKEY could make decisions, and AI was even solving algebra problems.

But the excitement faded. Computers weren't powerful enough, and AI systems struggled outside of controlled environments. By the 1970s and 1980s, AI hit a winter period, hype had outpaced reality, funding dried up, and the world moved on. For product managers today, this period serves as an important lesson: hype alone doesn't create value. If AI can't solve real-world problems, it won't last.

1990s–2000: machine learning changes the game

AI made a comeback in the 1990s and 2000s, not because of better algorithms, but because of more data and faster computers. Instead of manually programming rules, AI could now learn patterns from data. Here are a few more milestones from the 90s:

- 1997: IBM's Deep Blue defeats Garry Kasparov, proving that AI can master strategic decision-making.
- 1998: MIT introduces *Kismet*, an emotionally intelligent robot capable of detecting and responding to human emotions, an early step toward human-AI interaction.
- 1999: Sony launches *AIBO*, an AI-powered robotic pet that could learn and evolve over time, foreshadowing the personalization AI would bring to consumer products.

By the 2000s, AI started to enter everyday life

- 2002: The launch of *Roomba*, the first mass-produced autonomous vacuum cleaner, demonstrated AI's role in household automation.
- 2011: Apple's *Siri* introduced AI-powered voice assistance to the mainstream, making natural language interaction a daily reality.
- 2011: IBM's *Watson* won *Jeopardy!*, proving AI could process language and retrieve knowledge in real time, laying the foundation for modern AI assistants.
- 2014: Google's acquisition of DeepMind and the rise of AlphaGo demonstrated AI's ability to handle uncertainty, creativity, and strategic problem-solving, capabilities that would later prove crucial for generative models.
- 2014: Amazon *Alexa* popularized voice AI, redefining how users interact with technology.
- 2016: Microsoft's *Tay* chatbot, designed for social interaction, became a cautionary tale, demonstrating the risks of unmoderated AI learning and reinforcing the need for AI ethics.
- 2020: OpenAI's *GPT-3* marked a major leap in natural language generation, proving AI could generate human-like text at scale. This set the foundation for the GenAI explosion that followed.

AI was no longer confined to research labs; it was quietly transforming everyday products. Microsoft, Google, Facebook, and Amazon were embedding AI into search engines, recommendation systems, and advertising platforms. AI-powered virtual assistants, chatbots, and predictive analytics were becoming standard across industries.

Yet, while AI was advancing, it was still largely behind the scenes. GPT-3 and other early generative models were impressive, but their use was limited to developers and researchers. AI hadn't yet become an everyday tool for the average user.

That changed in late 2022, when generative AI became mainstream, putting world-class AI capabilities into the hands of millions.

November 2022: The Generative AI Revolution

The shift toward generative AI reached a tipping point in November 2022, with the release of *ChatGPT-3.5* by OpenAI. This wasn't just another AI milestone; it was the moment AI became mainstream.

For years, AI had quietly powered recommendation engines, search algorithms, and automation tools. But with ChatGPT-3.5, something changed. For the first time, AI wasn't just analyzing data or predicting patterns—it was writing, coding, designing, and generating entirely new content. It wasn't just a tool for automation anymore—it was a tool for creation.

And it didn't just impress AI researchers; it captured the world's imagination.

GenAI becomes the biggest conversation in every industry

I remember how, almost overnight, AI went from a background technology to the biggest conversation in tech, business, and beyond. Suddenly, professionals from all backgrounds including product managers, engineers, executives, and creatives were all asking the same questions:

- How can we integrate AI into our products?
- What impact will this have on my job?
- Is this just hype, or is it truly transformational?

Unlike previous AI breakthroughs that operated behind the scenes, this one was tangible. It wasn't just helping companies optimize supply chains or refining ad targeting, it was directly interacting with people. AI was no longer an invisible force working in the background, it was front and center, reshaping workflows, unlocking creativity, and redefining productivity.

This wasn't just another AI advancement; it felt like an industrial revolution unfolding in real time.

The steam engine of the mind

Silicon Valley pioneer Reid Hoffman compared generative AI to one of history's most transformative inventions:

"We should view Generative AI as a steam engine of the mind that promises to profoundly alter our professional and personal lives."—Reid Hoffman, Greylock Partners & Inflection AI

Just as the steam engine revolutionized industries by mechanizing production and expanding human capabilities, generative AI is poised to reshape how we create, communicate, and automate. This transformation isn't just about efficiency; it's about fundamentally changing the way we build and interact with technology. If traditional AI was about automation, generative AI is about amplification, scaling creativity, decision-making, and problem-solving beyond what was previously possible.

But such revolutions don't happen overnight. The question wasn't just about AI's potential, it was about why, after decades of research, this moment had arrived now.

What made 2022 the tipping point for Generative AI? Three forces driving Generative AI's rise

The Generative AI revolution that took off in November 2022 with ChatGPT's launch didn't happen overnight. It was the culmination of years of foundational breakthroughs across the AI industry. While OpenAI played a key role in advancing large language models, the broader AI ecosystem contributed three critical developments that made this moment possible:

Advances in compute power

The rise of cloud-based AI infrastructure and specialized AI chips such as Nvidia's A100 and H100 GPUs (graphics processing units) enabled the training of large-scale AI models at an unprecedented scale. These hardware advancements allowed researchers to push model sizes and training complexity to new limits, making AI systems faster, more powerful, and more efficient.

Explosion of training data

The internet's exponential growth spanning books, articles, social media, code repositories, and public datasets provided AI models with vast and diverse training material. This data abundance helped refine language comprehension and generative capabilities, allowing AI to generate human-like text with greater fluency, coherence, and contextual awareness.

Breakthroughs in model architecture

The transformer model, introduced by Google in 2017, revolutionized deep learning by allowing AI to process and generate text with a deeper understanding of context. OpenAI, along with other AI research teams, built upon this break-through to develop increasingly sophisticated models capable of longer, more co-herent conversations laying the groundwork for ChatGPT's viral adoption. And then, in November 2022, it all came together with ChatGPT-3.5. The world wasn't just introduced to Generative AI it was transformed by it overnight.

ChatGPT's meteoric rise and broader adoption

The pace of transformation was nothing short of astonishing. While past techno-logical revolutions unfolded over decades, ChatGPT's impact was immediate and undeniable.

ChatGPT didn't just showcase what generative AI can do, it set a new standard for real-world use. Within just two months, it reached 100 million active users. To put that into perspective, Instagram took 2.5 years, WhatsApp 3.5 years, and YouTube and Facebook nearly 4 years to hit the same milestone (Figure 2.2). This wasn't just an impressive statistic, it was a signal that something fundamental had shifted.

Path to 100 Million Users (stylized)

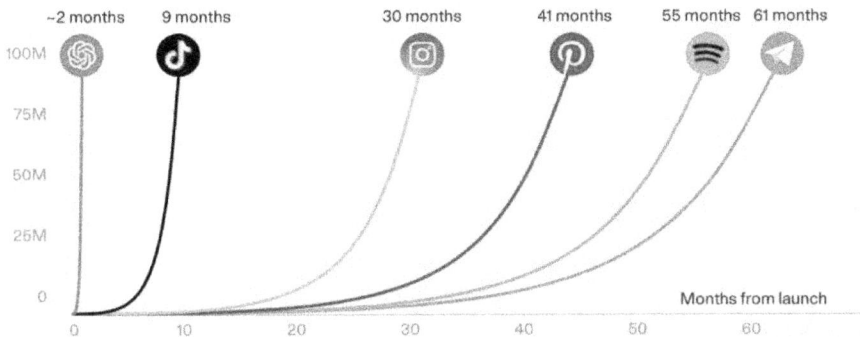

Figure 2.2. Path to 100 million users: a comparison of adoption speeds. *Source: Sequoia Capital.* *https://www.sequoiacap.com/article/generative-ai-act-two/*

ChatGPT wasn't an abstract AI innovation, it was something people could use, interact with, and immediately see the value in. Business leaders, software devel-opers, educators, content creators, and everyday consumers were experimenting, learning, and reshaping their workflows overnight.

For the first time, AI wasn't just a backend tool, it was in people's hands, reshap-ing how they worked and interacted with technology.

The race to catch up

As ChatGPT dominated headlines, the AI race intensified.

Anthropic, a company built on the premise of AI safety and accessibility, introduced Claude, a model designed to be both powerful and responsible. Initially, Claude could process around 9,000 words per minute, roughly the length of a short story. But within months, it improved dramatically, handling 100,000 words per minute, akin to the length of a novel. This leap underscored how quickly AI was evolving to meet the growing demands of content creation, research, and automation.

Meanwhile, Google was working to defend its search and AI dominance. It launched Bard, an AI-powered chatbot designed to enhance search experiences and user interactions. However, Bard's early reception was mixed, with critics pointing out its hallucinations and accuracy issues. Determined to regain momentum, Google introduced PaLM 2, an upgraded model that significantly improved contextual understanding and response quality. Later, Bard was rebranded as Gemini, marking Google's full-scale commitment to competing in the generative AI space.

Tech giants like Meta and Amazon weren't far behind. Meta focused on open-source AI, releasing Llama, a series of models designed for researchers and developers to build on. Amazon, on the other hand, invested heavily in AI infrastructure, integrating generative capabilities into AWS and expanding enterprise AI offerings.

In just a few months, what started as a breakthrough moment with ChatGPT had transformed into an all-out AI arms race. The competition wasn't just about building chatbots anymore, it was about shaping the future of AI-powered applications, automation, and knowledge work.

But while tech giants fought for dominance, another revolution was unfolding elsewhere—inside the enterprise.

From pilots to platforms: the enterprise turn into Generative AI

By early 2024, generative AI had started to move beyond experimental use and 65% of organizations were using GenAI tools in at least one part of their business, nearly twice as many as ten months earlier (McKinsey & Company 2024). Most of these efforts were still narrow in scope, but they signaled that adoption was no longer limited to tech pilots or isolated teams.

JPMorgan, for example, introduced a generative AI suite to support routine tasks across departments. Early reports suggested employees were saving several hours each week—time that could be redirected toward more judgment-driven work (Abrego 2025). It was a modest but meaningful step, showing how AI could fit into established workflows rather than overhaul them entirely.

That kind of incremental integration was common. By March 2025, only 1% of surveyed companies considered their GenAI rollouts "mature" (McKinsey & Company 2025). Most were still experimenting; some had launched initiatives without clear goals or measures of impact. But the organizations that were making progress tended to share a few patterns: they set specific objectives, tracked results, and gave senior leaders visibility into what was working and what wasn't.

While many challenges remain, the shift is underway. Organizations are no longer just trying to "get started" with generative AI, they're working to make it sustainable. Not as a headline-grabbing project, but as part of how the business operates. In that sense, generative AI is following a familiar pattern seen with other technologies: early excitement, uneven experimentation, and then a longer, quieter phase of operational refinement.

AI at work: professional adoption and cultural gaps

While many organizations were beginning to integrate generative AI into their operations, that shift wasn't unfolding evenly across the workforce. A November 2024 study conducted in Denmark offers one of the clearest snapshots of how AI tools, particularly ChatGPT, were actually being used at the individual level (Humlum and Vestergaard 2024).

Their survey focused on workers in eleven white-collar professions that were considered highly exposed to AI, including software development, marketing, customer support, and legal services. The findings painted a mixed picture. Developers and journalists were among the most active users. Others, such as financial advisers, teachers, and legal professionals, reported much lower levels of adoption, despite being aware of the tools and their potential. Figure 2.3 summarizes these differences by profession and task type.

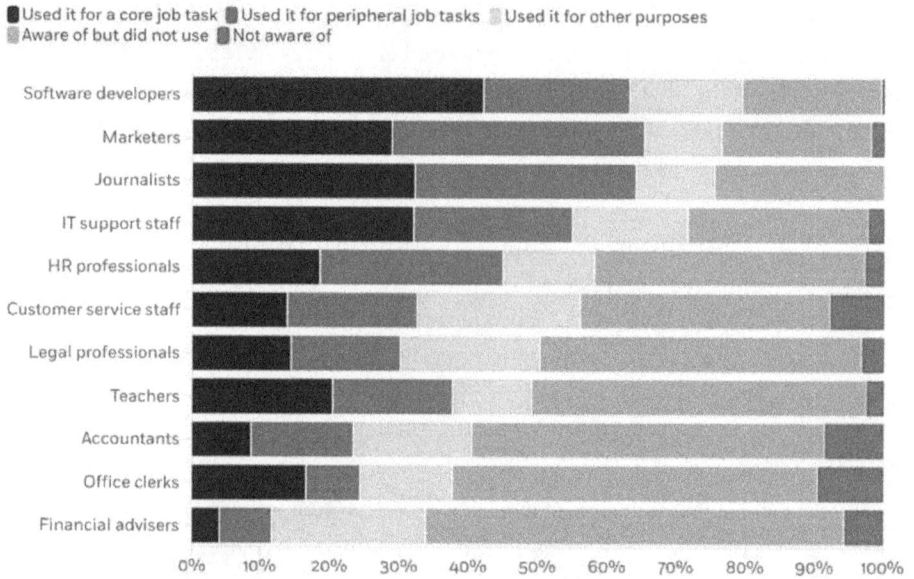

Figure 2.3. ChatGPT adoption rate by occupation and task type
Source: Chicago Booth Review. https://www.chicagobooth.edu/review/which-workers-are-embracing-ai

There were other divides, too. Younger workers and those earlier in their careers were more likely to use AI tools regularly. So were individuals with stronger academic records or higher pre-AI earnings. The study also uncovered a striking gender gap: women were 20 percentage points less likely than men to report using ChatGPT, even in similar roles. In many cases, women cited a lack of training. Men, on the other hand, more often pointed to data privacy concerns or company restrictions. Still, when offered additional resources, men were more likely to opt in.

Even among those who believed that ChatGPT could save time, sometimes by half, many didn't use it. The reasons weren't philosophical. Fewer than one in ten workers expressed concern about being replaced by AI. Instead, the barriers were more familiar: limited training, lack of support, or not knowing exactly how the tools applied to their day-to-day work.

These patterns suggest that availability alone doesn't guarantee adoption. Knowing that a tool exists isn't the same as knowing when and how to use it. For organizations, this gap between awareness and action presents a practical challenge, especially as they look to scale generative AI more broadly across roles and teams. Some industries moved quickly, embedding AI into daily operations, while others hesitated due to regulatory concerns, skill gaps, and trust issues. Fields like finance, legal, and healthcare, where accuracy and compliance are paramount, approached AI integration cautiously, weighing benefits against risks.

Beyond industries, AI adoption also varied by region, shaped by investment levels, government policies, and cultural attitudes toward automation. This raised a critical question: What factors were accelerating AI adoption in some regions while slowing it down in others? The answer lies in the broader GenAI market landscape, where investment trends, regulatory shifts, and competitive dynamics are shaping AI's future. But what does this landscape look like today?

GenAI Market Landscape

Understanding the GenAI market isn't just about tracking numbers, it's about identifying the forces driving this transformation and their impact on businesses, industries, and product managers worldwide. As companies race to integrate AI, three critical questions emerge:

- How big is the GenAI market today, and how fast is it growing?
- Which regions are leading in investment and adoption, and why?
- Who are the major players shaping this industry, and what competitive dynamics are at play?

Let's break it down.

Market size and growth trends

GenAI has rapidly evolved from an emerging technology to a significant economic force, transforming industries and redefining enterprise strategies. This acceleration is driven by breakthroughs in AI models, increased enterprise adoption, and surging investment in AI infrastructure.

According to recent forecasts from the International Data Corporation, global spending on AI—including software, hardware, and services—is expected to reach approximately $337 billion in 2025 and more than double to $749 billion by 2028, reflecting a compound annual growth rate of nearly 29% (Rooney 2025). While IDC's updated 2025 outlook did not break out a revised share for generative AI specifically, its earlier 2024 estimate projected that generative AI would account for roughly 32% of total AI spending by 2028, or around $202 billion (IDC 2024). The sharp upward revision in total AI spending underscores how the GenAI boom has significantly accelerated enterprise demand.

However, this growth is not just about increased spending, it reflects a fundamental shift in how businesses allocate AI budgets. Instead of traditional IT investments, such as software licenses and manual development, enterprises are prioritizing AI-driven automation, infrastructure, and scalable enterprise AI applications. Companies are moving from experimenting with AI to embedding it into core business processes, focusing on tools that automate repetitive tasks, enhance decision-making, and improve efficiency.

These global spending patterns signal more than just momentum, they reveal how organizations are beginning to prioritize AI not as an experimental add-on but as a foundational capability. Yet, the pace and shape of this transformation vary widely across geographies. Infrastructure maturity, regulatory clarity, and policy direction all influence how—and how fast—AI gets adopted. To understand where GenAI is heading, it helps to examine how different regions are approaching this shift.

Regional adoption patterns

As Generative AI adoption accelerates, different regions are taking unique approaches to innovation, regulation, and enterprise adoption. While North America leads in investment and commercialization, Asia-Pacific is rapidly expanding AI infrastructure, Europe is prioritizing ethical AI frameworks, and emerging markets are exploring AI for financial inclusion, agriculture, and public services.

This section explores how regional factors—government policies, investment patterns, and market readiness—are shaping AI adoption globally.

North America: the AI innovation hub

North America continues to lead the generative AI landscape, propelled by substantial venture capital investments, a robust AI research ecosystem, and widespread enterprise adoption. In Q1 2025, U.S.-based AI startups attracted a record $91.5 billion in venture capital, with 71% of that capital directed toward AI companies, underscoring the region's dominance in AI commercialization (Wall Street Journal 2025). Tech giants such as Microsoft, OpenAI, Google, Amazon, and Meta are at the forefront of foundational model development and enterprise-grade AI solutions, setting global benchmarks for AI adoption.

AI integration in North American enterprises is expanding rapidly, 60% of enterprise AI initiatives originate from the region (Gartner 2025). Organizations are embedding AI into customer interactions, business automation, and knowledge work, with a growing emphasis on AI-enhanced cloud services, cybersecurity, and enterprise applications.

However, regulatory challenges persist. While the U.S. government is exploring AI policies focused on safety, transparency, and responsible deployment, a lack of unified AI regulations creates uncertainty (World Economic Forum 2025a). AI governance frameworks are evolving, but companies must navigate varying compliance expectations.

Europe: the ethical AI powerhouse

Europe is establishing itself as a global leader in AI ethics and regulatory compliance, positioning responsible AI development at the center of its strategy.

The EU AI Act, the world's first comprehensive AI regulation, is shaping global AI policies by requiring strict adherence to safety, bias mitigation, and transparency standards. This regulatory-first approach is influencing how AI is deployed in highly regulated industries such as finance, healthcare, and insurance (EU Parliament 2025).

A European Commission Q1 2025 study found that 70% of European enterprises prioritize AI ethics and compliance, compared to 45% globally. This cautious, compliance-driven approach impacts AI adoption speed but fosters trustworthy AI ecosystems.

European AI startups and cloud providers such as DeepMind (UK), SAP AI (Germany), and Mistral AI (France) are gaining market share by offering General Data Protection Regulation (GDPR)-compliant and bias-aware AI models. While AI experimentation may be slower due to regulations, Europe is setting global AI safety standards that could influence future AI laws worldwide.

Asia-Pacific: The AI acceleration zone

Asia-Pacific (APAC) is experiencing rapid AI transformation, fueled by government-led AI initiatives, industrial automation, and smart city investments.

China, India, Japan, and South Korea are at the forefront, with China's AI spending growing at ~35% CAGR, positioning it as the second-largest AI market after the U.S (IDC 2024; World Economic Forum 2025a). The region benefits from strong AI infrastructure investments, domestic model development, and AI-powered automation in manufacturing, finance, and logistics (World Economic Forum 2025a).

China leads in AI-driven manufacturing, fintech, and e-commerce automation, with state-sponsored investments supporting tech giants such as Alibaba Cloud, Huawei, and Tencent AI (World Economic Forum 2025a). India is emerging as a global hub for enterprise AI solutions, with firms like Infosys, TCS, and Wipro integrating AI into IT services, SaaS platforms, and business process automation (Economic Times 2025).

Meanwhile, Japan and South Korea are applying AI to robotics, logistics, and semiconductor innovation, aligning AI growth with their advanced industrial capabilities. South Korea, for instance, plans to secure 10,000 high-performance GPUs for national AI infrastructure (Reuters 2025). Across the region, governments are proactively shaping the AI landscape. China's sovereign AI strategy, South Korea's AI innovation zones, and Singapore's AI-first digital economy policies are accelerating adoption—solidifying Asia-Pacific as one of the fastest-growing AI economies globally (World Economic Forum 2025a; World Economic Forum 2024).

Latin America & Africa: AI for emerging markets

While AI adoption in Latin America and Africa is at an earlier stage, both regions are experiencing significant momentum in key AI-driven sectors.

- Financial Inclusion: AI-powered digital payments, fraud detection, and micro-lending platforms are increasing financial access in underserved communities (CAF 2025).
- Agriculture & Climate Tech: AI is being utilized for crop monitoring, irrigation optimization, and climate forecasting, enhancing food security and promoting sustainable farming practices (Reuters 2025a).
- Healthcare & Public Services: AI chatbots and predictive analytics are improving healthcare accessibility and automating government services (Etori et al. 2023).

Cities like São Paulo (Brazil), Nairobi (Kenya), and Lagos (Nigeria) are emerging as regional AI innovation hubs, attracting investments in AI-driven fintech, healthcare, and e-commerce solutions (Brookings Institution 2023).

Despite these advancements, challenges remain, including limited cloud access, AI infrastructure gaps, and talent shortages. However, global AI partnerships and investment collaborations are helping to bridge these gaps, enabling the regions to leapfrog into AI adoption in critical sectors (Brookings Institution 2023).

What this means for product managers

Understanding regional AI adoption is critical for product managers designing and scaling AI-powered products.

- AI product localization—AI-powered platforms must adapt to regional regulations, data privacy laws, and language preferences.
- Go-to-market strategies—Expanding into Europe requires compliance-first AI, while APAC markets demand AI tailored for automation and industrial applications.
- Regulatory monitoring—Keeping up with evolving AI policies (EU AI Act, U.S. AI safety regulations, China's AI governance) is essential for risk management.

As AI continues reshaping global markets, product managers must anticipate shifting adoption trends, compliance challenges, and competitive landscapes. These insights will be crucial for defining AI business models, partnerships, and scaling strategies—topics we will explore next.

Now that we've explored investment trends and AI's evolving role in enterprises, let's look at who is driving this transformation. The next section examines the major players shaping generative AI, from foundation model providers to enterprise AI platforms, while also highlighting the companies leading AI agent innovation.

Major Players in the Generative AI Market

The rise of generative AI has triggered a competitive race among technology giants, AI research labs, cloud providers, and open-source communities. But this competition is no longer just about building the largest and most powerful AI models, it's about shaping AI ecosystems that seamlessly integrate models, agents, and enterprise applications.

We've moved past the era of standalone models. The next frontier is multi-agent AI architectures, where AI systems collaborate to perform specialized tasks, retrieve external knowledge, and automate workflows. As companies shift toward multi-model AI strategies, success isn't just about the raw power of a model it's about how effectively AI can be orchestrated, deployed, and embedded into real-world applications.

To understand the major players shaping this transformation, we need to look beyond just who builds the models and explore the entire AI value chain. This section provides a high-level overview of the companies driving the AI industry forward, showing how different players fit into the evolving AI value chain. While we'll dive deeper into the technical and product implications of these layers in later chapters, this overview will help contextualize the key organizations shaping the future of generative AI.

Note: If you're encountering terms that feel unfamiliar, don't worry. Chapter 3 will walk through the foundational concepts behind GenAI, and you'll also find a detailed glossary in the Appendix A for quick reference. Feel free to read forward or loop back as needed, this book is designed to support you at every step of your AI product journey.

Now, let's break it down.

Layer 1: AI infrastructure (the foundation of AI compute)

AI doesn't exist in a vacuum, it requires massive computational power. Every breakthrough in AI, from ChatGPT to Deep Research (autonomous agent), depends on specialized hardware and cloud infrastructure to train, deploy, and operate models at scale. Without compute power, there are no large-scale AI models, no inferencing, and no AI agents.

Companies that dominate AI compute don't just provide hardware; they shape the cost, accessibility, and innovation speed of AI itself. As of 2025, the AI compute ecosystem spans both cloud and on-premises deployments.

While hyperscale public clouds remain central to model training and large-scale serving, on-prem and edge clusters are gaining traction—especially for agentic AI systems embedded in regulated industries, robotics platforms, and latency-sensitive applications. This ecosystem can be understood through three key layers:

- AI accelerators—Specialized chips (GPUs, Tensor Processing Units, or TPUs, and AI processors) designed to handle AI training and inference.
- Cloud and on-prem AI platforms—The compute substrate where AI workloads run, whether in public clouds, private data centers, or hybrid edge setups.
- Networking & storage—The backbone that enables AI models to process vast amounts of data efficiently and at scale.

These layers determine how fast AI progresses, who gets access to cutting-edge capabilities, and how affordable AI-powered solutions can be for businesses.

- Nvidia: The backbone of modern AI, providing the most powerful GPUs used for training AI models.

- Google, Microsoft, Amazon Web Services (AWS): Competing with their own AI chips (TPUs, Maia, Trainium) to reduce reliance on Nvidia.

- Intel & AMD: Developing next-gen AI chips to challenge the dominance of GPUs.

- Cloud giants (Azure AI Foundry, AWS Bedrock, Google Vertex AI): Powering cloud-based AI workloads, while increasingly supporting hybrid and on-prem deployments for use cases requiring data locality, low latency, or regulatory compliance.

Why It Matters: The AI arms race isn't just about building better models, it's about who owns the compute layer. AI is only as powerful as the infrastructure supporting it, and these companies dictate who gets access to the most advanced AI capabilities.

Layer 2: AI intelligence (foundation models)

Foundation models are the core intelligence driving AI applications. Companies today aren't just choosing one model, they're using multiple AI models in parallel, selecting the right tool for the job. The foundation models landscape is broad and evolving (as shown in Figure 2.4), with significant diversity in model capabilities.

High-performing generalist models such as GPT-4, Claude Opus, and Gemini 1.5 Pro (positioned in the top-right quadrant) are optimized for broad reasoning across multiple domains. In contrast, smaller models like Mistral 7B and Llama 3.2 3B (positioned with fewer parameters) prioritize efficiency, cost, and on-device deployment.

Additionally, industry-specific models (not shown in Figure 2.4) are emerging, often built on top of these foundation models to serve specialized use cases. This trade-off between power, cost, and specialization shapes how businesses choose the right AI model for their needs.

100 MMLU

GPT o1*

Claude 3 Opus

89.8 = human expert ———— Emie 4.0 ———————————— Claude 3.5 Sonnet* ——— GPT o1 pro*

GPT-4o* DeepSeek-V3

Claude 3.5 Sonnet*

GPT-4o* Llama 3.1 405B Grok-2 n2.5 Llama 3.3

80

GPT-4 Classic Gemini 1.5 Pro

U-PaLM Nemotron-4-340B

Claude 2 Claude 2.1 Gemini 2.0

Falcon 180B

70+ IDEAL ———— Titan —————————— Arctic ——— Gemini-1.5

Pixtral-12b

LLaMA-65B Llama 2 NeMo

Chinchilla InternLM2

Emie 3.5 K2 Llama 3.2 3B

60

Skywork-13B MAP-Neo

Gopher Granite Minitron-4B

Baichuan 2 Pile-T5 OLMoE-1B-7B

Galactica JetMoE-8B

Atlas Griffin

40 GPT-3 HLAT

RWKV-v5 EagleX

BLOOM Command-R

UL2 20B BloombergGPT Hawk

GPT-NeoX RWKV-v5 Eagle 7B Rene

Mistral 7B

EXAONE 3.0

RoBERTa OpenELM Device Jun 24

Mamba

20

AMD-Llama-135m

| pre-2022 | 2022 | 2023 | 2024 | 2025 |

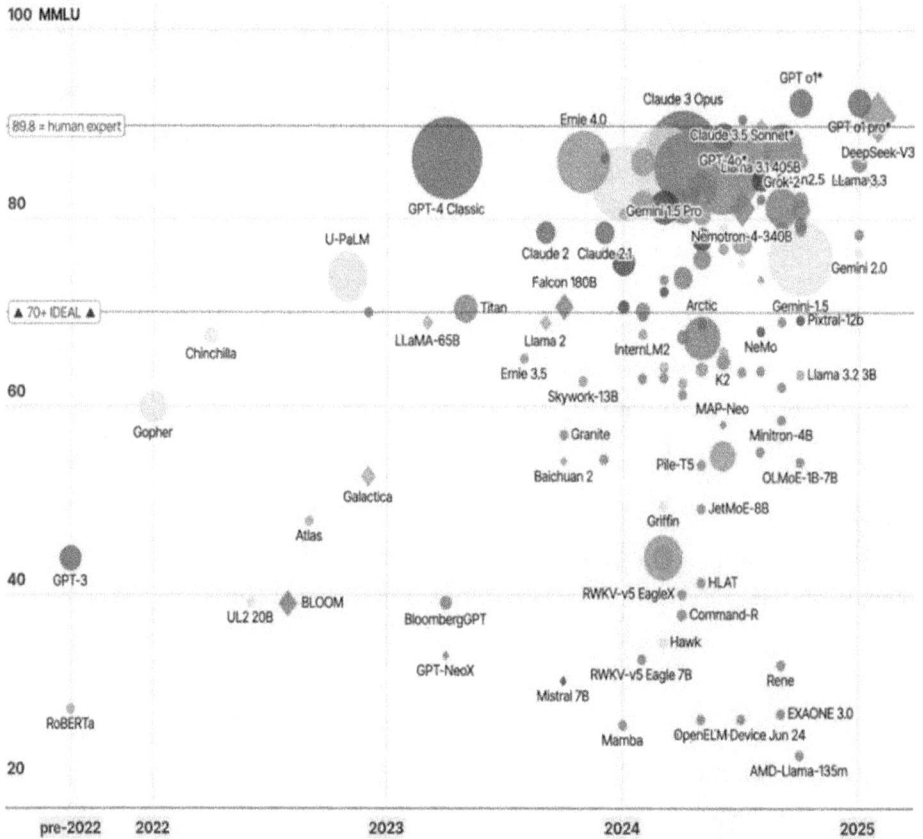

Figure 2.4. The evolving landscape of foundation models, ranked by capabilities and sized by
billion parameters used for training.
Source: Information is Beautiful. https://lifearchitect.ai/models/

But model performance alone isn't the only factor in AI adoption. Companies evaluate several key considerations when selecting a model:

- Access & deployment—Do they want an API-based proprietary model or a self-hosted open-source model?
- Customization & control—Can they fine-tune the model, or must they rely on the provider's updates?
- Security & compliance—Are there regulatory concerns that favor open vs. closed models?

Given these primary considerations for companies in how they choose AI models, let's look at two broad categories based on how they are accessed, modified, and controlled:

- Proprietary models *(closed-source, API-driven, scalable but limited customization)*
- Open-source models *(publicly available weights, customizable, with the option to self-host or use via managed APIs)*

This classification is crucial because it determines how companies integrate AI into their tech stack, how much control they have over model behavior, and the long-term cost implications.

Proprietary models (closed-source, API-driven)

These models are fully owned and controlled by companies that do not publicly release their model weights. Instead, they are offered as API services, businesses can use them but cannot modify or self-host them.

Advantages: high performance, enterprise-grade security, no infrastructure setup required.
Limitations: Limited customization, potential vendor lock-in, high costs.

Major Players:

- OpenAI: Known for its deep integration with enterprise cloud platforms, OpenAI delivers scalable AI solutions for businesses.
- Google DeepMind: Specializes in multimodal AI, embedding intelligence across productivity tools, search, and enterprise applications.
- Anthropic: Focused on AI safety and responsible development, making it a preferred provider in regulated industries such as healthcare and finance.
- Cohere: Designed for domain-specific AI with a strong foundation in retrieval-augmented generation (RAG) for knowledge-intensive applications.
- AI21 Labs: Develops language models optimized for enterprise applications.

Open-source models: a spectrum of access and control

Today's AI model ecosystem isn't split neatly into "open" and "closed" it's a spectrum, ranging from fully open-source projects to hybrid offerings that blend openness with commercial services. Understanding this spectrum is essential for product teams choosing between control, customization, support, and ease of deployment.

Fully open-source models (self-hosted, customizable)

These models provide public weights and architecture under permissive licenses like MIT or Apache 2.0, enabling complete control. Businesses can download, fine-tune, and deploy these models on their own infrastructure, managing everything from scaling to compliance.

Advantages: Full control, no vendor lock-in, cost-effective for long-term AI strategies.
Limitations: Requires significant technical expertise and operational ownership.

Major Players:

- Meta (Llama series): When self-hosted, these models offer broad customization under open licenses.
- Microsoft (Phi series): Open-source model family released under the MIT License, optimized for reasoning tasks.
- Google DeepMind: Released the Gemma family of open models, providing open weights and permissive terms that allow for broad use, including commercial applications. Gemma models are designed to be lightweight and efficient, suitable for deployment across various platforms.
- Mistral AI: Compact, high-performance models (e.g., Mistral 7B, Mixtral) released under Apache 2.0.
- DeepSeek AI: Strengthens domestic AI capabilities, focusing on AI independence and sovereignty.
- Hugging Face: A platform that hosts and maintains thousands of open-source AI models.
- OpenAI (gpt-oss series): The newly released gpt-oss-120B and gpt-oss-20B are fully open-weight reasoning models available under Apache 2.0, marking OpenAI's first open-weight release since GPT-2.

Hybrid-access models (blending openness and enterprise readiness)

In hybrid models, the base model is open or accessible, but the provider offers enterprise-grade value-adds—such as hosting, fine-tuned APIs, SLAs, and premium tools.

This category of models are increasingly popular as businesses seek both customization and support at scale. Think of it like Linux vs. Red Hat: the core is open, but a commercial wrapper makes it easier to deploy, maintain, and scale.

Advantages: Lower setup burden, faster deployment, production-ready reliability. Limitations: Some features may be gated, and control is traded for convenience.

Major Players:

- Meta's Llama (via Azure or AWS): Open weights, but available as managed APIs with enterprise support.
- Mistral (Mixtral + Hosted APIs): Public weights plus premium, production-ready deployment options.
- Stability AI: Open models like Stable Diffusion, complemented by enterprise services and hosted infrastructure.
- AI21 Labs: Offers open-access models and API-based commercial solutions.

Note: Companies like Meta and Mistral appear in both categories because it depends on how their models are used. When users self-host the models, it's an open-source use case. When they access the same models through commercial APIs or hosted platforms, it shifts into the hybrid territory.

Why It Matters: The future of AI is multi-model. Companies are no longer tied to a single provider or architecture, they're combining proprietary APIs with open-source models, whether self-hosted or accessed through hybrid solutions. This evolving landscape demands better orchestration tools and sharper decisions about control, customization, and infrastructure ownership.

Layer 3: AI logic & execution layer

This layer powers the "thinking and doing" inside GenAI applications, including agentic AI—where decisions are made, tasks are sequenced, and external knowledge is integrated in real time. It's what transforms AI from a content generator into a capable collaborator. Whether you're chaining together foundation models with task-specific models or enabling an autonomous agent to decide what to do next, this is the layer where intelligence becomes action.

This layer provides the essential building blocks for the following:

- Planning & task coordination: Orchestration frameworks like LangChain or Semantic Kernel help break big goals into smaller steps. They guide the AI to decide what to do next based on the task, user input, or context.

- Memory & context retention: To keep track of user preferences or task progress, AI systems need memory. Orchestration frameworks help manage what should be remembered, when to retrieve it, and how to maintain a coherent experience across steps or sessions.

- Tool invocation logic: When AI systems need to take action like sending a message or pulling live data, developers define logic in orchestration frameworks to decide *which* tool to use, *when* to use it, and *how* to process the outcome or errors.

- Retrieval logic (RAG): When the AI needs real-time or internal knowledge, orchestration frameworks route the query to a retrieval system (like a vector database), fetch relevant content, and feed it into the model's prompt to improve accuracy and relevance.

- Multi-agent collaboration: In more complex workflows, multiple agents may each handle a specific part of a task. Orchestration frameworks manage how these agents communicate, share progress, and hand off tasks smoothly like teammates working toward a shared goal.

Major Players

- Core agent frameworks: LangChain, Semantic Kernel (Microsoft), CrewAI, Autogen (Microsoft), LlamaIndex (supports agent frameworks)

- Retrieval systems (RAG): Pinecone, Weaviate, ChromaDB, Azure AI Search (Cloud-based search with integrated vector capabilities).

- Tool invocation & API interfaces: LangChain Tools, Function Calling APIs (OpenAI, Claude), custom API connectors

- Memory handling: LangMem, MemGPT, Memo, Zep (purpose-built memory modules); LangChain Memory; Redis and PostgreSQL (commonly used for custom or persistent memory stores)

Why It Matters: This is the backbone of agentic AI—without orchestration, AI agents can't reason, retrieve knowledge, or automate workflows. It is the layer that allows agents to move from simple tasks, to complex problem solving.

Layer 4: AI hosting & serving (where AI models & agents are deployed)

Once models are trained and orchestrated, they must be deployed for real-world use at scale. This layer provides the infrastructure and services for hosting, and serving. It also encompasses customization capabilities, allowing for the adaptation of models and agents to specific domain needs.

AI hosting has traditionally focused on model deployment, but as AI agents gain traction, agent hosting and serving is emerging as a distinct category. Many of these platforms have already evolved into AI model marketplaces, where businesses can buy, fine-tune, and deploy pre-trained AI models. The same is being seen for the agents as well.

Major Players:

- Cloud AI model hosting & serving: Azure AI Foundry, OpenAI API, Google Vertex AI, AWS Bedrock, Hugging Face Inference Endpoints
- Agent hosting & serving (emerging market): OpenAI Assistants API, Microsoft Azure AI Agent Service, AWS Bedrock, Mistral Agents API
- Inference optimization companies and technologies: Nvidia, Qualcomm, Intel
- AI marketplaces (integrated into hosting platforms): Hugging Face Hub, Azure AI Foundry Catalog, AWS Bedrock, Google Model Garden

Why It Matters:

- AI inferencing costs are a major concern—companies are optimizing for scalability, efficiency, and performance to reduce infrastructure expenses.
- Agent hosting & serving is becoming its own market, enabling businesses to run and deploy autonomous AI agents alongside traditional AI models.
- Enterprises are demanding dedicated AI hosting solutions that go beyond API-based access, allowing for fine-tuned AI deployments with better control over security, latency, and costs.

Layer 5: AI governance, observability & responsible AI

As AI adoption grows, ensuring fairness, transparency, and accountability is crucial—especially as AI systems become more autonomous. This layer covers how AI is monitored in production (observability), how it complies with regulatory and organizational policies (governance), and how it stays ethical and fair in its behavior (responsible AI).

While today's systems focus on governing model performance, bias, and compliance (e.g., GDPR, HIPAA), future agentic AI will require oversight of autonomous decisions, tool usage, and multi-agent interactions.

Major Players:

- AI performance monitoring: Weights & Biases, Arize AI, WhyLabs
- Bias detection & fairness: Microsoft Responsible AI, AI Fairness 360 (IBM), Holistic AI
- Explainability & governance tools: Arthur AI, Google Vertex AI Explainability, Credo AI, Truera, Fiddler AI

Why It Matters: Without strong governance, AI can become a black box, prone to bias, unfairness, and unintended harm. Responsible AI ensures that models today—and agents in the future—are trustworthy, accountable, and aligned with ethical principles.

Layer 6: AI application and interface (where AI meets the user)

This layer captures the surface of the GenAI stack—where users interact with AI through apps, copilots, and agent-driven workflows. It includes AI-augmented enterprise tools, AI-native products, and platforms that let teams build or embed AI experiences. This is where AI becomes tangible, transforming how work gets done, decisions are made, and creativity flows.

Major Players:

- Enterprise AI copilots: Microsoft 365 Copilot, Google Workspace Duet AI, Salesforce Einstein GPT
- AI-native applications: Notion AI, Canva AI, Figma (AI features), Perplexity AI (conversational search), ChatGPT (OpenAI's flagship consumer app)
- No-code/low-code AI builders: Zapier AI, Retool AI, Bubble AI, Microsoft Copilot Studio (for building domain-specific copilots and chat agents)

Why It Matters: The next wave of AI adoption will be driven by AI-native applications and agent-driven experiences that seamlessly integrate AI into everyday workflows.

As GenAI adoption accelerates, it's reshaping the global economy through technological innovation, industry transformation, and new ways of creating value. The companies leading AI innovation are not only advancing the technology but also finding innovative ways to monetize it, driving economic growth. We'll explore these emerging business models in detail in Chapter 7, but for now, let's examine how GenAI is transforming industries and societies at scale.

Economic and Industry Impact of Generative AI

By now, we've explored where GenAI meets the user—whether through copilots inside everyday tools, AI-native apps like Perplexity or Notion AI, or builder platforms like Microsoft Copilot Studio and Azure AI Foundry that help teams bring their own agentic workflows to life. These interfaces aren't just new features, they're changing how people work, learn, and make decisions.

But the changes don't stop at the user interface. Underneath, entire industries are starting to shift. Unlike past tech waves like cloud or mobile, which reshaped businesses over years, GenAI is compressing that change into quarters—or even sprints. Industries like healthcare, finance, education, and manufacturing aren't just adopting AI, they're reorganizing around it.

As a GenAI PM, your work won't live in isolation. The tools you build will ripple into hiring plans, workflows, and business models. To understand that momentum, let's zoom out first to the economic forces enabling this shift, then into how it's unfolding industry by industry.

The Economic Forces Driving Gen AI Adoption

AI is no longer confined to research labs or innovation teams, it's impacting GDP, productivity, and business efficiency across every sector. Generative AI has grown at an unprecedented rate, powered by advances in foundation models and the rise of AI-native applications.

But beyond adoption metrics and product integration, GenAI is fundamentally altering economic trajectories: reshaping how work is done, how industries are structured, and how value is created and captured in the global economy.

Analysts estimate that generative AI alone could add between $2.6 trillion to $4.4 trillion annually to the global economy, with use cases spanning from customer operations and software engineering to R&D and marketing (Chui et al. 2023). AI is no longer just a tool for automation, it is a catalyst for economic transformation, changing how businesses operate, how work gets done, and how industries evolve.

Two primary forces are driving this economic shift. First, GenAI is unlocking new levels of efficiency by reducing operational costs, accelerating task completion, and enhancing decision-making. These productivity gains are not limited to technical teams, they're spreading across functions like legal, HR, marketing, and customer service.

Second, GenAI is fundamentally reshaping how work gets done, shifting human roles from execution to orchestration. This shift from workers doing tasks to managing AI agents that do tasks is already starting to change the structure of labor markets. Let's unpack the first of these forces: the productivity and cost advantages that are prompting organizations to adopt GenAI at scale.

Cost reduction and productivity gains

One of the biggest economic forces behind AI adoption is productivity acceleration. Enterprises that have successfully integrated AI into their operations are reporting upto 40% efficiency gains in knowledge work tasks such as software development, data analysis, customer service, and marketing (McKinsey & Company, 2025).

AI-driven automation is enabling businesses to do the following:

- Reduce operational costs by automating repetitive and manual tasks.
- Increase output with AI-assisted content generation, coding, and workflow automation.
- Optimize decision-making through AI-powered analytics, forecasting, and risk assessment.

However, AI's economic impact is not just about efficiency, it is about creating entirely new ways to monetize technology, structure businesses, and generate competitive advantage. With cloud AI platforms making GenAI widely accessible, AI is no longer a luxury for tech companies, it is becoming a core component of every business strategy. Companies that fail to adapt to this shift risk being left behind in an AI-first economy.

Workforce transformation: a consequence and driver of GenAI adoption

As aspiring GenAI product managers, you're stepping into a world where companies are grappling with real labor challenges—too few workers, rising costs, and pressure to stay competitive.

Take healthcare: hospitals are facing a nurse shortage, with some regions reporting vacancy rates of 20%, forcing reliance on expensive temporary staff. In customer service, call centers struggle to hire enough agents, with turnover rates reaching 30% annually (McKinsey & Company 2025). Logistics firms report up to 15% of warehouse roles unfilled in key markets like the U.S. (Volberda 2025). These labor gaps strain operations, so companies are turning to GenAI to automate tasks and ease the load.

According to the World Economic Forum, AI could displace 92 million jobs by 2030, especially in clerical, administrative, and repetitive white-collar roles while also creating 170 million new ones (World Economic Forum 2025). This isn't just about job loss; it's about how work is changing. Instead of grinding through tasks, workers are now overseeing AI workflows, training models, and making decisions informed by AI systems (McKinsey & Company 2025).

Here's how it plays out: when companies embed GenAI into operations, they start seeing real outcomes. Up to 46% of administrative tasks, like filing reports or scheduling, can be automated, allowing teams to accomplish more with fewer people. Tasks get done faster: customer queries are resolved in seconds instead of minutes. Productivity jumps by as much as 40% in software development roles (McKinsey & Company 2025). These gains create a feedback loop: success drives further investment in GenAI, which leads to more operational efficiency and leaner, higher-performing teams.

That's why workforce transformation isn't just a side effect of GenAI adoption, it's a major driver. Companies aren't just reacting to change; they're pursuing it. By helping close labor gaps and enable new ways of working, GenAI offers both a survival strategy and a competitive edge.

For you as product managers, this means building AI solutions that solve real workforce pain points like automating hospital scheduling or simplifying customer support. But it also means designing tools that uplift workers, not just replace them.

Of course, the shift isn't frictionless. Reskilling workers to collaborate with AI takes time, money, and thoughtful planning (World Economic Forum 2025). If we're not careful, we risk leaving people behind. That's why your job includes designing tools that are intuitive, supportive, and inclusive.

While these shifts in the workforce are significant, they're just one part of the story. GenAI isn't just changing how people work, it's transforming how entire industries operate. From healthcare and software development to finance and retail, the ripple effects of GenAI adoption are visible across every major sector. Let's take a closer look.

GenAI Across Industries: Sector-Specific Shifts You Need to Know

In the last section, we looked at how GenAI is reshaping the workforce, changing what people do and how companies operate. But that transformation doesn't look the same everywhere.

The way generative AI shows up in a hospital isn't the same as how it reshapes a retail supply chain or a trading floor. Each industry is adapting in its own way, driven by its own pain points, constraints, and opportunities. As a GenAI product manager, understanding these differences is essential. The best solutions don't just apply AI they solve real problems for real sectors. So, let's look at how GenAI is playing out across some of the world's largest industries and what that means for the products you'll build.

Healthcare: scaling expertise, easing burnout

Healthcare systems around the world are under pressure not just from staff shortages, but from rising patient volumes, documentation demands, and increasingly complex care protocols. Generative AI isn't replacing doctors. It's helping them focus on what matters most: treating patients.

Hospitals in the U.S., Japan, and Europe are piloting GenAI tools that cut documentation time by 50–70%, automating tasks like clinical summaries, follow-up scheduling, and even patient communication (Harvard Gazette 2025). These time savings are freeing up hours each week for frontline providers.

At the same time, GenAI is accelerating medical research. AI assistants are helping scientists process huge datasets, identify promising drug targets, and even generate novel compounds. In some trials, drug discovery timelines have dropped from 6–7 years to under two (Zhang, Yang, and Yang 2025).

These tools aren't just copilots anymore. In specific scenarios, they're starting to operate autonomously flagging anomalies in imaging data, prioritizing patient risks, or surfacing next-best actions. But the challenge isn't just building more capable tools, it's trusted ones. In healthcare, GenAI must be transparent, reliable, and easy to integrate into clinical routines that are already stretched thin.

As a product manager, your goal isn't only to maximize efficiency. It's to design tools that support better outcomes without adding friction or creating new risks. Think of assistants that simplify workflows, not systems that overwhelm with complexity.

Software development: from copilots to autonomous agents

Software development is undergoing a significant transformation. Generative AI is evolving from merely assisting developers to taking on more autonomous roles. Tools like GitHub Copilot have paved the way, but newer systems are now capable of identifying bugs, writing pull requests, running tests, and deploying updates with minimal human intervention.

According to Sequoia Capital, teams leveraging generative AI development agents have reduced time-to-market by up to 60%. This shift allows developers to focus more on system design and user experience, while AI handles routine coding tasks (Sequoia Capital 2024).

For product managers, this means rethinking the development pipeline. Integrating AI tools into existing workflows requires careful consideration of version control, continuous integration, and deployment processes. Ensuring that AI-generated code meets quality and security standards is paramount.

Finance: from forecasting to intelligent execution

Finance has long been a proving ground for automation—credit scoring, fraud detection, and market forecasting were early wins for machine learning. But generative AI is moving the industry beyond prediction into intelligent execution. JPMorgan Chase, for instance, has deployed a proprietary GenAI platform called LLM Suite, which supports over 600 use cases across the firm. It helps draft investment memos, summarize research, generate client communications, and even analyze earnings calls to surface trading signals—all within seconds (Saeedy 2025). This shift isn't just about speed, it's about surfacing insight from complex, unstructured data.

On the frontlines, AI-driven assistants are enhancing customer service. These systems don't just respond to scripted queries; they guide customers through nuanced processes like selecting mortgage options or understanding investment risk. Behind the scenes, compliance teams use GenAI to automate regulatory research, flag anomalies in transactions, and streamline Know Your Customer (KYC) workflows.

But in a domain where the stakes are high and regulations tight, trust is non-negotiable. A model that can't explain its decision—or gives the wrong one—can cost millions or lead to legal exposure. That's why financial GenAI must be auditable, interpretable, and human-overridable by design.

If you're building for this space, think beyond automation. Your role is to craft AI systems that illuminate—not obscure—decisions. Make room for transparency. Show the reasoning. And always keep humans in the loop.

Retail: reimagining the customer experience

Retailers are using GenAI to reinvent how products are marketed, merchandised, and sold. But the shift isn't just digital storefronts with AI chat. It's about personalized journeys that stretch across search, selection, purchase, and service.

Take fashion: global brands are using GenAI to generate product descriptions in dozens of languages, automatically update catalog metadata, and even forecast demand based on social media trends. In e-commerce, AI agents are helping shoppers find what they need, not just through filters, but by asking natural-language questions (Standish et al. 2024).

On the backend, AI is improving inventory management and supply chain responsiveness. GenAI models can analyze point-of-sale data, logistics timelines, and even weather forecasts to suggest restocking decisions. This reduces overstock, minimizes waste, and improves margins. For retail PMs, one opportunity lies in agent-led commerce—AI systems that act like virtual stylists or buying assistants. The challenge? Ensuring these experiences are trustworthy, brand-aligned, and non-intrusive. A smart shopping assistant that overwhelms or misguides users can do more harm than good.

GenAI in retail is ultimately about relevance, offering the right product to the right person, at the right time. Your job is to make that invisible intelligence feel seamless.

Manufacturing: from smart factories to self-optimizing systems

Manufacturing has long embraced automation, but GenAI is unlocking a new phase: factories that learn, adapt, and optimize themselves.

Instead of just automating repetitive processes, manufacturers are now using AI agents to analyze production data, identify bottlenecks, and dynamically adjust machine settings in real time. Companies like Siemens and Rockwell are integrating GenAI into industrial IoT platforms, enabling predictive maintenance, quality control, and demand-responsive workflows (Siemens 2025).

For instance, GenAI models can scan video feeds to detect product defects, flag anomalies in sensor data, and generate reports summarizing root causes. Some plants have already reported up to 25% improvement in overall equipment effectiveness (OEE) and 30% reductions in unplanned downtime by integrating AI-based diagnostics (McKinsey 2025).

But the benefits go beyond the factory floor. AI copilots are being used by engineers to simulate production environments, test changes, and auto-generate documentation—freeing up time for higher-level problem solving.

For GenAI PMs in this space, the opportunity is to bridge AI and operations. Your products must integrate seamlessly with existing industrial systems, work reliably in real-time settings, and surface actionable insights without overwhelming users.

Impact on other industries

While sectors like healthcare, finance, and software development are leading GenAI adoption, the momentum is building elsewhere too. In government, public agencies are using GenAI to automate call center responses, draft policy documents, and support multilingual services, helping overworked departments deliver more with fewer resources. Logistics and transportation firms are deploying AI copilots to optimize routes, manage fleet schedules, and anticipate supply chain disruptions before they happen. In the energy sector, AI models now assist in grid forecasting, climate risk modeling, and predictive maintenance, especially in renewables, where variability requires constant recalibration.

Legal teams, once buried under case files, are using GenAI to summarize contracts, synthesize precedents, and accelerate e-discovery processes, reducing weeks of work to days. In education, schools and edtech platforms are integrating GenAI to provide personalized learning plans, automate feedback, and support students with real-time tutoring—making differentiated instruction more scalable than ever.

And in media and entertainment, studios are experimenting with AI for everything from editing and visual effects to script drafts and content personalization—tools that speed up production cycles while giving creators a head start.

Final thought

Across these sectors, GenAI isn't replacing core expertise, it's amplifying it. For product managers, this means looking for friction-heavy workflows, data-rich environments, or compliance-heavy processes where intelligent automation can make a measurable impact. Even in slower-moving industries, the foundational shift is underway. The race isn't about who adopts AI first but who integrates it meaningfully into how work gets done.

That said, the journey toward AI-native products comes with its own headwinds. Regulations are tightening as governments move quickly to shape how AI is built and used—from the EU AI Act to U.S. executive orders and China's evolving compliance frameworks (OECD 2025; World Economic Forum 2025c). Meanwhile, high compute costs can limit experimentation, especially for smaller players (IDC 2025), and model brittleness—hallucinations, bias, or lack of transparency—remains a barrier in fields where trust and accountability are non-negotiable.

But these aren't just engineering challenges, they're product challenges. They sit at the intersection of user trust, technical feasibility, and responsible innovation. And solving them is what sets great GenAI product managers apart.

This moment isn't just another digital transformation. It's a foundational shift—one where the very idea of what software can do, how work gets done, and what teams look like is being redefined. And the people who will shape that future are already on the front lines.

One profession that's feeling this shift most directly is product management. As AI moves from augmentation to fully AI-native ecosystems, PMs must rethink their role, tools, and mindset. Traditional frameworks built around feature roadmaps and quarterly iterations are giving way to a new paradigm one defined by real-time orchestration, multi-agent collaboration, and continuous learning.

GenAI's Impact on Product Management

This shift reframes the very essence of product management. What was once a craft centered on roadmaps, feature delivery, and incremental optimization is now becoming a discipline of orchestrating intelligence—across models, tools, data, and users.

In the GenAI era, AI isn't a layer you sprinkle on top. It's the substrate—the foundation around which products are imagined, built, and evolved. This isn't an incremental shift. It's systemic, transforming how user value is created, how systems adapt, and how trust is maintained across the stack. You're no longer just managing features. You're designing systems that reason, improve, and sometimes surprise. And that means a new mindset, one that embraces uncertainty, prioritizes experimentation, and views model behavior as part of the user experience.

For aspiring GenAI PMs, the lesson is clear: You can't just "add AI" to existing workflows. You must build around it, embedding intelligence at the core of the product experience. GenAI PM doesn't just ship software. They productize intelligence turning raw model capabilities into usable, trustworthy, and valuable experiences.

To meet this new paradigm, product managers must go deeper than ever before. It's no longer enough to own a backlog or define a user journey. In GenAI, PMs must understand how models behave, how they're hosted, how they interact with tools and users and how those systems evolve over time. That's why we're witnessing the rise of a new kind of product leader.

The Rise of the Full-Stack AI Product Manager

Today, GenAI isn't about layering interfaces over pre-trained models. It's about shaping behavior across the entire stack from infrastructure to models to orchestration to user interfaces and adapting that behavior as the system learns from user interactions.

This shift is giving rise to a new kind of product manager: the full-stack AI PM. These aren't just feature owners, they're system thinkers. They don't just ship features, they orchestrate intelligence.

To thrive in this role, you'll need fluency across four essential layers:

- Model behavior: How does the model respond to different inputs? What are its limitations? How does it learn over time?
- Tooling & orchestration: What tools or agents control the flow of logic, retrieve information, or trigger actions behind the scenes?
- Infrastructure: Where and how is the model hosted? How do you balance cost, latency, and scalability?
- User experience & outcomes: How do users interact with the system? How do you ensure trust, clarity, and real-world impact?

In GenAI, your product isn't static. It's a living system one that evolves with every interaction. And as a full-stack PM, your job is to guide that evolution shaping systems that think, adapt, and earn trust as they grow always aligned with human needs and values.

The Shift from Feature-Centric to AI-first Thinking

When smartphones first emerged, some companies simply resized websites for smaller screens. But the real winners Uber, Instagram, Snapchat didn't retrofit the past. They reimagined experiences from the ground up, based on how people behave on mobile. The shift wasn't just technical, it was behavioral, experiential, and strategic.

Generative AI demands the same kind of rethinking. This isn't just about adding AI to existing workflows. It's about reimagining those workflows and sometimes the product itself through an AI-first lens. PMs who treat AI as a sidecar will miss the deeper transformation: AI-first thinking reshapes how we define value, what users expect, and how systems evolve. So, the core question becomes: how would you design your product if AI wasn't a feature but the foundation?

AI-first product thinking means asking:

- What human frustrations can AI meaningfully reduce?
- What dynamic, adaptive experiences can we now unlock?
- How do we co-create with AI?
- How do we embed trust and feedback from the start?

In an AI-first world, the product isn't static. It's interactive, generative, and ever-learning. It's not just about enhancing the experience; it's about transforming it. That shift doesn't start with the tech. It starts with how product managers think.

New UX Mandates: Trust, Transparency, and Control

AI doesn't just change product functionality; it changes user psychology. Unlike traditional software, AI outputs are non-deterministic. They vary. They surprise. Without strong UX guardrails, they confuse or worse, erode trust.

Today's PMs must design for three functions:

- Explainability: Showing how decisions or outputs are formed (e.g., surfacing sources, revealing steps).
- User control: Giving users clear ways to intervene, override, or refine AI behavior.
- Failure recovery: Making mistakes understandable and recoverable—instead of silent or confusing failures.

A vivid real-world example came with Google's AI Overviews, which rolled out broadly to the public in early 2024. Intended to augment search results with AI-generated summaries, the feature quickly drew backlash when users encountered hallucinations, including a now-famous suggestion that glue could help cheese stick to pizza. The issue wasn't just an odd error it was a breach of user trust at internet scale (BBC 2024).

In GenAI products, trust isn't a layer you add it's the foundation you build on. Without explainability, control, and recovery, even brilliant capabilities collapse under the weight of broken trust.

Beyond Static Features: Toward Adaptive Experiences

Traditional products were rule-based: fixed UIs, predefined logic, consistent outputs. AI-first products behave differently:

- They adapt to user behavior in real time (e.g., Notion AI suggesting edits mid-writing).
- They generate personalized, context-aware outputs (e.g., ChatGPT, Adobe Firefly).
- They predict intent before a user explicitly acts (e.g., Gmail Smart Compose).

Take Microsoft 365 Copilot as an example. Instead of simply embedding AI-powered search into Word and Excel, Microsoft reimagined productivity software itself—transforming documents and spreadsheets into collaborative AI-powered workspaces where AI drafts content, summarizes meetings, and even suggests actions. This wasn't just an "AI add-on", it was a fundamental shift in how people work with documents.

Building for GenAI means building dynamic, behavior-driven experiences, not just workflows.

AI as a Co-Builder Across the Product Lifecycle

GenAI is no longer just an accelerant, it's a creative collaborator embedded across every stage of the product lifecycle.

In ideation and research, it synthesizes customer insights, identifies market gaps, and forecasts emerging trends, just as Spotify uses AI to detect micro-trends in music genres. During design and prototyping, it auto-generates mockups, interaction flows, and even UX copy, collapsing the time from idea to prototype (Figma AI lets designers go from prompt to layout in seconds).

In development and testing, it accelerates coding, automates QA, and predicts performance issues before they escalate—like GitHub Copilot, which offers real-time code suggestions that boost developer velocity. And in post-launch evolution, AI personalizes user experiences, tunes systems in real time, and analyzes feedback loops to keep products responsive and adaptive.

In the GenAI era, PMs aren't just collaborating with engineers they're collaborating with adaptive systems that learn, respond, and co-create at every layer of product development.

Redefining Success Metrics for AI Products

Traditional metrics like DAU, MAU, and retention tell only part of the story in AI-powered products. In GenAI, success isn't just about usage, it's about *quality*, *adaptability*, and *trust* over time.

AI-first product managers must track new dimensions of performance:

- AI engagement: How often do users interact with AI-generated outputs? What percentage of key workflows are AI-assisted?
- Model quality: How well does the model understand context, avoid hallucinations, and improve with feedback?
- Inference efficiency: Are we balancing accuracy, latency, and cost—especially in GPU-intensive workloads?
- Human-AI collaboration: Do users trust the AI? Are they adopting, correcting, or relying on its outputs meaningfully?

Measuring GenAI products requires a shift in mindset: you're not just optimizing for usage, you're optimizing for learning, alignment, and responsible behavior at scale.

Toward an Agentic Future: PMs Guiding Autonomous Systems

We've already seen AI agents show up across this chapter—from optimizing production lines in manufacturing to reshaping how roles evolve in the workplace—and even earlier in Chapter 1, where they signaled a shift in what product managers now design. But what exactly makes an AI system agentic? And why does it matter for product managers?

As GenAI evolves, so does its capacity for autonomy. We're moving from AI assistants—designed to support humans through suggestions and completions—to AI agents: goal-driven entities capable of reasoning, taking action, and adapting without step-by-step instructions.

These agents can:

- Execute workflows end-to-end (e.g., scheduling, triaging support tickets)
- Adapt strategies in real-time (e.g., portfolio optimization in finance)
- Collaborate across agent networks to complete multi-step objectives

In sectors like finance, agentic systems are already analyzing trades, adjusting strategies, and responding to market conditions often without human intervention.

For PMs, this changes the game:

- How do you monitor and guide evolving agent behavior to ensure it stays aligned with user intent?
- How do you design escalation paths, define guardrails, and correct drift over time?
- How do you measure success when your product continues learning, adapting, and changing after it ships?

The future of product management isn't about launching static software. It's about orchestrating intelligent, autonomous systems that evolve responsibly, always anchored in human goals, values, and trust.

Final Thought: The AI-First PM Mindset

The best product managers of the GenAI era won't just ship features, they'll architect adaptive, agentic systems, embed trust through transparency and resilience, and partner with AI as a true co-creator. They'll guide intelligent agents that evolve with every interaction—earning trust, delivering value, and staying grounded in human goals.

If you embrace AI not as an add-on, but as a co-builder—
If you design for behavior, not just outcomes—
If you lead with trust, not just functionality—

You won't just survive the GenAI revolution.
You'll shape what comes next.

Before you can lead AI products, though, you need to understand what powers them.

In the next chapter, we'll break down how generative AI actually works so you can make smarter product decisions, speak the language of your engineering partners, and build with confidence in an AI-first world.

Chapter 3: Foundations of Generative and Agentic AI

Generative AI is no longer just a futuristic concept, it's a transformative force reshaping industries, products, and how we work. While early AI systems focused on rule-based automation or pattern recognition, today's models can generate original content, reason through problems, and even take autonomous actions.

For aspiring AI product managers, understanding how generative AI works, how it differs from traditional AI, and how it's evolving into agentic systems is critical. Whether you're designing AI-powered applications, fine-tuning models, or enabling intelligent agents, you need a solid foundation in how these systems are structured.

Before we unpack how GenAI models create, generate, and act, it's important to understand where they sit in the broader AI landscape.

The AI Landscape: From Rules to Autonomy

Artificial Intelligence isn't a single technology; it's a layered ecosystem of evolving capabilities (see Figure 3.1). From simple rule-based systems to fully autonomous agents, each layer represents a step forward in intelligence, complexity, and potential impact. Let's walk through it.

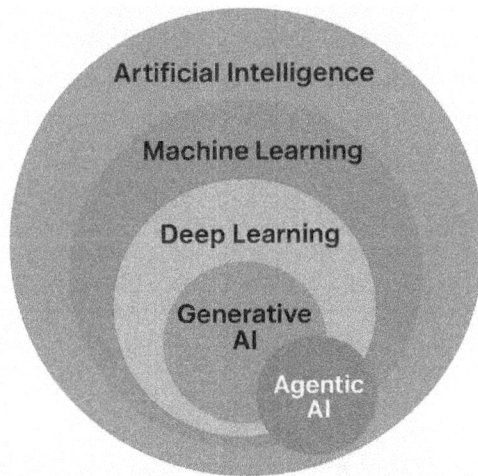

Figure 3.1: How GenAI and Agentic AI fit into the AI landscape
© 2025 by the author of *Zero to GenAI Product Leader.*

Artificial Intelligence (AI)

At the broadest level, AI refers to systems designed to mimic aspects of human intelligence such as reasoning, decision-making, or problem-solving. Early AI systems were rule-based: they followed hard-coded instructions to make decisions in predictable environments. Over time, AI began incorporating methods that could adapt, learn, and improve based on data.

This shift led to the rise of machine learning.

Machine Learning (ML)

Machine learning is a subset of AI that enables systems to learn from data rather than rely on rigid rules. Instead of explicitly programming every scenario, engineers feed the system examples. The machine identifies patterns and uses them to make predictions or decisions.

There are three foundational approaches: supervised learning, which learns from labeled examples (like email spam detection); unsupervised learning, which identifies patterns in unlabeled data (like customer segmentation); and reinforcement learning, which improves performance through feedback (as in game-playing bots).

These ML systems now power many product features we take for granted such as personalized recommendations, fraud detection, or churn prediction. They're reliable tools for well-defined problems with lots of data.

Neural networks—ML models inspired by how the brain processes information—are the bridge that connects traditional learning algorithms to deep learning architectures.

Deep Learning (DL)

Deep learning takes things a step further. It's a type of machine learning built on artificial neural networks, layered structures inspired by how the brain processes information. Deep learning models excel at handling complex, unstructured data like text, images, and audio. These models made major advances in language translation, image recognition, and speech-to-text systems. Unlike traditional ML algorithms that needed carefully selected inputs, deep learning models could learn features directly from raw data, making them ideal for building more sophisticated AI capabilities.

By allowing models to grasp nuance and context in data, deep learning became the foundation for breakthroughs in language, vision, and generative modeling. In many ways, it laid the groundwork for the next wave: generative AI.

Generative AI (GenAI)

Generative AI refers to systems that can create new content—text, images, music, code, and more. Unlike traditional models that categorize or predict, GenAI models generate something novel based on the patterns they've learned. They're built on deep learning architectures, especially transformer models like GPT, LLaMA, or Claude.

This shift enabled breakthroughs like ChatGPT drafting emails, GitHub Copilot writing code, and tools like Midjourney generating digital art from text prompts. Generative AI has redefined what's possible in product experiences not just reacting to users but co-creating with them.

It's a powerful capability but it also raises new product challenges around trust, output quality, and responsible use. As GenAI systems become embedded into everything from copilots to creative tools, product managers must balance innovation with control.

Agentic AI

Agentic AI builds on generative AI but shifts the focus from creation to execution. These systems combine generative capabilities with memory, planning, tool use, and feedback mechanisms. They don't just generate outputs, they follow through on tasks.

Most agentic systems today are built on top of GenAI models. For example, they may use a language model like GPT-4 to understand a request, plan steps to complete it, retrieve information using tools or APIs, and adjust behavior based on results. Frameworks like LangChain, Semantic Kernel, and AutoGPT make this orchestration possible.

However, not all agentic AI systems are generative. Some use traditional planning techniques, rule-based logic, or reinforcement learning to operate in structured environments without ever generating content. That's why agentic AI is best understood as overlapping with generative AI. Most examples today are generative, but not all need to be.

Think of the difference this way:

- Generative AI might write a great draft.
- Agentic AI figures out what to write, finds supporting sources, edits the draft, and sends it when you're ready.

For product managers, this evolution marks a shift from designing static AI responses to crafting full task completion flows where AI behaves more like a collaborator in the user's workflow than a standalone tool.

Why this matters:

Understanding this hierarchy helps product managers choose the right AI technology for different use cases (e.g., ML for predictions, GenAI for content creation); collaborate effectively with engineers and data scientists; and assess feasibility & constraints (e.g., compute-intensive deep learning models vs. lightweight ML models).

Understanding this layered structure of AI—how each tier builds upon the last—helps product managers make informed choices. Whether you're designing features powered by ML or exploring autonomous workflows with agentic systems, this hierarchy helps you evaluate tradeoffs, feasibility, and technical complexity.

More importantly, it sharpens your instincts. You'll know when a simple classification model is enough—and when the opportunity calls for something that plans, adapts, or generates. Generative AI changed how we build. Agentic AI is changing how products behave, shifting AI from a tool to a collaborator. But to build confidently with these systems, you need to understand how they actually work not just what they produce, but how they learn, reason, and respond.

Let's start by demystifying generative AI.

How Generative AI Works

Generative AI might feel like magic, but under the hood, it follows a structured learning and reasoning process. From ingesting massive datasets to generating nuanced, human-like responses, each stage plays a role in making GenAI smart, scalable, and useful for real-world products.

As a PM, you don't need to build these models but understanding the key components will help you make smarter product decisions, collaborate better with engineers, and design experiences that feel intuitive and trustworthy.

Let's walk through the foundational building blocks from training to inference in the order they typically occur in a GenAI system's lifecycle.

Training Data and Model Learning

Everything starts with data.

Generative AI models like GPT-4 or Claude are trained on enormous amounts of text, code, images, and more. This data is used to teach the model how language works, how ideas connect, and how to respond in contextually appropriate ways.

During this stage, the model is not memorizing specific responses, it's learning *patterns*. Just like we learn to speak by hearing conversations, GenAI learns by predicting the next word (or token) based on what it's seen before.

For example, models like GPT are trained on datasets such as Common Crawl (a large snapshot of the web), Wikipedia, and large collections of books. This helps them generate human-like responses, even when they've never seen your exact input before.

Why this matters for PMs:

Understanding what a model has been trained on helps you anticipate its strengths, gaps, and potential biases and design products that handle those responsibly.

Tokenization—The Input Breakdown

Before the model can make sense of language, it has to break it into digestible chunks called *tokens*. These are the smallest units the model processes like syllables or word fragments. These tokens can be whole words, parts of words, or even individual characters, depending on the model.

Instead of predicting an entire word or sentence at once, the model works by predicting one token at a time, step by step. It uses each predicted token as context to generate the next one, which is how it forms complete responses. For example, the word *"unbelievable"* might be broken into the tokens: *"un"*, *"believ"*, and *"able"*

Tokenization affects how long prompts can be, how responses are generated, and how usage is priced (many GenAI platforms charge per token). A basic understanding helps you design better interactions and estimate costs more accurately.

Neural Network Architectures—The Power Behind Generative AI

Modern generative AI is built using a type of deep learning architecture called the *transformer*, which was introduced in 2017. It changed how AI models work by allowing them to read and understand entire chunks of text or images all at once, rather than word-by-word or step-by-step like older systems.

A key reason transformers are so powerful is something called *self-attention*. This means the model can look at every word in a sentence and figure out which words are most important, based on the context. For example, in the sentence *"The trophy didn't fit in the suitcase because it was too big,"* the model uses self-attention to understand that *"it"* refers to *"trophy,"* not *"suitcase."* This helps AI generate much more natural and accurate responses.

Transformers also rely on something called *parameters* these are the settings the model learns during training. You can think of parameters like the AI's memory of everything it has seen and learned. Modern generative AI models like GPT-4 are trained on huge amounts of text and have billions (or even trillions) of parameters, which allow them to generate responses that seem thoughtful and creative.

Why this matters for PMs:

While you don't need to understand the math behind transformers, knowing they allow deep context awareness helps you design better flows for summarization, Q&A, or multi-turn conversation experiences.

Embeddings—Teaching the Model Meaning

Once tokens are processed, the model represents them as embeddings—mathematical vectors that capture their meaning based on context.

Think of embeddings as how the model "understands" relationships. Words like *"dog"* and *"puppy"* live close together in this vector space, while *"dog"* and *"democracy"* are far apart.

Why this matters for PMs:

Embeddings power features like personalization, semantic search, and recommendation systems. They're a critical part of making AI feel more relevant and helpful to users.

Vectors—Connecting the Model to Knowledge

Embeddings become even more powerful when stored and queried through vector databases like Pinecone or Weaviate. This enables retrieval techniques, where the model pulls real-time information before responding. A common method is retrieval-augmented generation, which combines the model with external data sources in real-time. For example, RAG helps a support bot answer accurately by pulling from a knowledge base, ensuring responses are grounded in real data. It often works with retrieval tools like vector databases (e.g., Pinecone) to fetch real-time information, a key component in agentic systems like those using *agentic RAG*, which combine retrieval with autonomous actions. Advanced techniques like GraphRAG extend this by using graph-based knowledge for relational reasoning, such as analyzing connections in research data.

Why this matters for PMs:

Vectors and retrieval techniques like RAG enable your product to deliver precise, context-aware responses, but they require access to reliable data sources, something to plan for when designing features like customer support or research tools.

Fine-Tuning and Adaptation

After the base model is trained, it can be *fine-tuned* for specific tasks or industries, meaning it's retrained on smaller, domain-specific datasets to improve performance in niche areas. For example, OpenAI's Codex is fine-tuned for code generation, enabling tools like GitHub Copilot; DALL·E is fine-tuned to generate images from text descriptions.

Fine-tuning makes the model more relevant and accurate in specific domains such as customer service, legal document summarization, or creative writing.

Why this matters for PMs:
Fine-tuning can improve relevance and accuracy, but it's expensive and requires specialized data and compute. In many cases, product teams may opt for lighter alternatives, such as the following:

- Prompt engineering is used to craft better inputs to guide model behavior

- Retrieval techniques involve combining the model with external data sources in real time, also known as retrieval-augmented generation, as discussed above.

Inference and Content Generation

Once a model is trained, it enters the inference phase the process of generating responses based on new user input. When a user interacts with a GenAI model like typing a prompt into ChatGPT the model doesn't pull a response from a database. Instead, it uses what it has learned to predict the most likely next token, one by one, until the response is complete.

Because this process is probabilistic (there's more than one possible "next word"), the outputs can be creative and flexible but also inconsistent. It can also generate incorrect or fictional information, a phenomenon known as hallucination.

Why this matters for PMs:

PMs must account for the fact that GenAI outputs may vary and occasionally be inaccurate. Good product design must include UX guardrails and feedback mechanisms to maintain trust and usability and consider fallback options for when things go off track.

Each component in this chain—training data, tokenization, embeddings, vectors, fine-tuning, and inference—shapes how a GenAI model responds to user input. Together, they define the experience of interacting with AI systems that feel intelligent, helpful, and even creative.

But as magical as it may seem, generative AI still comes with real constraints. It doesn't know what happened after its training cut-off. It can't remember you from one session to the next. And it sometimes produces results that sound right but aren't. These limitations become more visible—and more critical—when we try to use GenAI in real-world products.

Let's briefly look at where today's GenAI systems struggle and why those struggles are driving the next evolution of AI.

The Limitations of Generative AI

Generative AI can be impressive at first glance—writing code, drafting content, or responding to questions with ease. But as product managers start building with it, its boundaries become clear. There are real constraints—technical, behavioral, and contextual—that shape what it can reliably deliver in practice.

These models stand out because they've been trained on massive datasets and can mimic fluency, creativity, and reasoning. But they don't *understand* the world. They don't *know* you. And they certainly don't adapt without guidance.

That's because most GenAI systems are still inherently limited in five key ways:

First, they *forget*. Most large language models (LLMs) are stateless by design. That means they treat each prompt like a blank slate unless memory systems are manually added around them. Ask ChatGPT for vacation ideas today, and tomorrow it won't remember where you said you wanted to go unless it's part of a longer session or you've built persistent memory into the app.

Second, they *hallucinate*. If you've ever seen an AI confidently cite a non-existent report or invent a fact about your company, you've witnessed this firsthand. These models generate based on patterns, not truth. They'll make things up sometimes subtly, sometimes egregiously especially when prompted with uncommon or nuanced queries.

Third, they *don't know what's happening right now*. A model trained on data from 2023 has no idea what happened in March 2025 unless you give it access to that information. Without retrieval systems, GenAI can't access your current product roadmap, your customer's recent support ticket, or your latest pricing sheet.

Then there's *cost and speed*. Running inference on large models, especially with long or multi-turn interactions, can be expensive and slow. For some consumer applications, that's manageable. For enterprise-scale tools or real-time systems, it becomes a real constraint.

And finally, they're *inconsistent*. Ask the same question twice and you may get slightly different responses. That's not a bug, it's part of how these models generate creative, probabilistic outputs. But in certain workflows like legal summaries or compliance tools variability can be a deal-breaker.

Why this matters for PMs:

59

These aren't just technical gaps, they're design challenges. As product managers, we're not just shipping models; we're shaping user experiences. And those experiences break down when the AI forgets something important, hallucinates confidently, or fails to act on what it just generated.

The good news? Many of these gaps are solvable not by replacing the model, but by building around it: adding memory, connecting tools, orchestrating steps, and grounding responses in real-time data.

This is the shift we're now living through: from generating content to enabling action. From GenAI to something more autonomous, more goal-driven, and more useful in the flow of real work.

Let's explore what happens when AI doesn't just respond but plans, decides, and executes. That's the world of Agentic AI.

From Generation to Autonomy: The Rise of Agentic AI

In Chapter 1 and earlier in this chapter, we introduced agentic AI as a new design pattern—AI systems that don't just respond, but reason, plan, and act. Now, having explored where generative AI falls short in real-world settings, let's take a closer look at how product teams are starting to address those gaps by building agentic systems.

Agentic AI isn't a replacement for GenAI, it's an evolution. As mentioned earlier, most agentic systems today still rely on foundation models at their core. But what makes them agentic is how they're designed: with structured layers like memory, planning, tool use, and feedback loops that turn raw generation into reliable execution.

We've already seen how GenAI systems can impress in isolated tasks. But when the goal is completing a workflow—across systems, adapting to changes, or coordinating with humans—product teams need more than just outputs. They need orchestration.

That's the heart of agentic design: surrounding the model with the right scaffolding so it can actually *do* something useful with what it generates.

This is the real unlock for product teams: You're no longer just designing prompt/response flows. You're enabling goal completion, crafting AI systems that act with purpose, respond to context, and integrate seamlessly into user workflows. Instead of prompting a foundation model every time, imagine giving it a high-level task like: *"Schedule a team meeting for next week, gather agenda items, and notify attendees."*

A generative model might draft an email. An agent figures out what to write, checks calendars, drafts the message, sends it—and follows up if plans change. These capabilities don't come from the model alone—they come from the structured system around it.

Let's now unpack what makes this possible, the core building blocks of agentic AI.

Core Building Blocks of Agentic AI

If generative AI is the engine that understands and generates, agentic AI is the full vehicle—capable of planning the route, steering, adjusting, and even fixing itself along the way. So, how do we go from passive generation to intelligent action?

Agentic AI systems are made possible by layering several components on top of a core foundation model. These components work together to give AI systems autonomy, adaptability, and purpose. As a product manager, you don't need to implement these yourself—but understanding what they do will help you define what your product *should* do.

Earlier, we explored the six-layer AI value chain from infrastructure and model development to orchestration, user interfaces, and governance. That framework helped us understand how companies and technologies fit into the broader GenAI ecosystem. Now, let's zoom in. Inside any single AI system, especially one that behaves autonomously, a more focused architecture comes into play (see Figure 3.2). At the heart of every agentic AI system are three essential functional components that work in concert to bring intelligence to life:

- Model (the brain): The reasoning and content generation engine
- Orchestration (the decision-maker): The system that enables planning, memory, and workflow execution
- Tools (the hands & eyes): The components that allow AI to interact with the world, take action, and retrieve fresh information

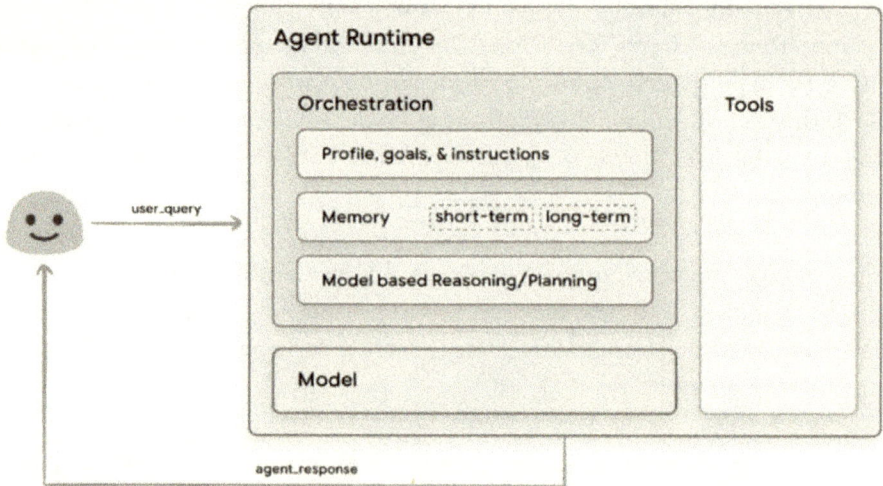

Figure 3.2. Functional components of agentic AI.
Source: *Google Agents Whitepaper.* https://ppc.land/content/files/2025/01/Newwhitepaper_Agents2.pdf

These building blocks map directly to what we previously covered: *models* align with the model layer, *orchestration* corresponds to the AI logic & execution layer, and *tools* represent the external systems that are connected through that orchestration—whether it's APIs, databases, or business applications. Next, let's break them down.

GenAI model: the brain of the agent

Agentic AI starts with a GenAI model (LLMs like GPT-4, Claude, Gemini) as the core reasoning engine. It's what makes the agent sound fluent and capable when responding. This model is responsible for the following:

- Understanding user input
- Reasoning through problems
- Generating responses and structured outputs

But while the model is powerful, it has limitations:

- It hallucinates (i.e., generates plausible but incorrect information).
- It lacks memory, it forgets what happened in previous sessions.
- It has no real-world access, it can't fetch live data, run calculations, or perform actions on its own.

Why this matters for PMs:

- The model is the foundation of an AI agent but not the entire system.
- PMs must choose the right model (proprietary vs. open-source) based on cost, accuracy, and customization needs.
- Understanding LLM constraints (hallucination, token limits, real-time access) helps design better AI workflows.

Since models alone aren't enough to create intelligent agents, we introduce orchestration, the layer that gives AI its decision-making, memory, and structured workflow execution.

Orchestration: the decision-maker of the agent

Orchestration is what turns a static model into a dynamic, task-executing agent. It enables the following:

- Memory (so the agent retains past interactions)
- Planning (so it can break down tasks into structured steps)
- Workflow execution (so it follows logical loops instead of answering blindly)

Memory: retaining context across time

Since GenAI models are stateless meaning, they don't retain past interactions once a conversation ends every new query is treated as an independent request.

Agentic AI systems introduce memory so the AI can recall relevant details across interactions whether it's remembering a user's preferences, past conversations, or progress on a task. Memory can be one of two types:

- Short-term (retained within a single session)
- Long-term (persisting across days, projects, or user interactions)

You might be asking, wait—*doesn't ChatGPT already have memory?* It might feel that way, especially when it responds based on earlier messages in the same chat. But that's context awareness, not true memory.

Here's the difference:

- Context awareness: Tools like ChatGPT can refer back to earlier messages but only within a limited "context window" (a set number of tokens).

- No persistent memory: Start a new chat, and the AI forgets everything you discussed before unless specialized APIs or settings are used.

Agentic AI takes this further: It stores information across sessions, enabling AI to remember what you like, what you've done, and where you left off just like a real assistant. This unlocks long-term planning, continuity, and more personalized execution.

Why this matters for PMs:

Memory makes AI feel more intelligent by enabling personalization, continuity, and contextual awareness. However, storing memory adds complexity, requiring considerations around privacy, security, data freshness, and retrieval accuracy. PMs must define what should be remembered, for how long, and when to discard outdated context.

Planning: breaking down complex goals

For an AI agent to complete tasks, it needs to do more than just respond, it needs to plan. That means taking a high-level instruction (like *"draft a proposal"*) and figuring out the necessary steps:

1. Understand the goal: What does "draft a proposal" mean in this context?
2. Break it down into steps: Research the topic, outline the content, write a first draft, format it.
3. Decide the sequence: Some steps may need iteration or dependencies.

How do AI agents plan?

To complete tasks effectively, AI agents need a planning mechanism. Instead of responding instantly to a request, they break the goal into logical steps, just as a human would when solving a complex problem.

Chain-of-thought (CoT) reasoning: thinking step by step

This approach helps an AI solve problems logically by explicitly reasoning through intermediate steps just like a person showing their work on a math problem.

For example, imagine you ask an AI, "What's 35×47?" Without reasoning, it would simply output 1,645, but with chain-of-thought reasoning, the AI shows its thinking process:

- Step 1: $30 \times 47 = 1,410$

- Step 2: $5 \times 47 = 235$
- Step 3: Add both results → 1,645

CoT reasoning is particularly useful for math problems, logical reasoning, and complex decision-making where breaking a problem into steps improves accuracy.

ReAct (reason + act): Combining thinking with execution

While CoT reasoning focuses on step-by-step thinking, ReAct takes it a step further—it combines reasoning with actions.

Instead of just thinking through a problem, the agent can also interact with tools, retrieve information, or execute actions as part of the reasoning process.

For example, imagine an AI agent helping a user book a flight. Using ReAct, the AI follows these steps:

1. Reason: "To book a flight, I need the user's destination and preferred date."
2. Act: The AI asks the user for missing details.
3. Reason: "Now I have the details. I need to find available flights."
4. Act: The AI calls an API to fetch real-time flight availability.
5. Reason: "Here are the top three options based on price and timing."
6. Act: The AI presents the best options to the user.

ReAct enables agents to interact with external tools and retrieve live data while reasoning, making them more dynamic and capable in real-world applications. Earlier AI systems relied on task graphs to map workflows in a rigid, rule-based manner. While effective for static automation (like traditional bots), modern AI agents use ReAct and chain-of-thought (CoT) reasoning for more flexible, real-time decision-making.

Why this matters for PMs:

Planning is the backbone of any useful agent. Without planning, an agent might stop halfway, produce incoherent results, or skip critical steps, making it unreliable for real-world tasks.

Managing the workflows

An agent isn't just a one-and-done system. It often operates in loops: plan → act → reflect → revise. This flow needs coordination. Orchestration tools like LangChain, AutoGen, and Semantic Kernel coordinate the logic behind those steps, manage memory, and ensure the agent doesn't go off the rails. Think of this as the operating system that keeps the agent from getting lost or repeating itself.

Why this matters for PMs:

Orchestration ensures AI stays on track and doesn't repeat or contradict itself. PMs should collaborate with engineers to define how user input, memory, tools, and retrieval should interact to create a seamless user experience. In addition, PMs shape the workflow design, deciding when the AI should act autonomously vs. when it should defer to a human (escalation paths).

Reinforcement learning: learning from experience

While most AI agents follow predefined rules and workflows, some emerging systems can learn from experience and adjust their actions over time. Unlike traditional training, which happens before deployment, *reinforcement learning* (RL) allows AI to improve during real-world use. This is still in its early stage for most GenAI applications but is an exciting direction for self-optimizing agents in dynamic environments.

Why this matters for PMs:

Most products won't need this right away but knowing what's possible helps you ask smarter questions about your agent's adaptability and future roadmap.

Tool use: expanding agent capabilities

Generative AI models are great at generating content, but they can't take action or retrieve live information on their own. They don't know how to send an email, book a meeting, run a calculation, or fetch real-time data unless they're connected to external tools.

Tool use is what transforms a model into a capable agent. It allows the AI to interact with the world by calling APIs, querying databases, updating spreadsheets, sending messages, or performing calculations. For example, it can allow agents to do the following

- Take actions (e.g., sending emails, scheduling meetings)
- Fetch real-time data (e.g., checking live stock prices)
- Query internal knowledge (e.g., retrieving company policies)
- Perform calculations (e.g., using external computation engines)

Tools bridge the gap between a passive content generator and an interactive AI agent that can execute tasks, fetch relevant data, and automate workflows.

Think of it like this: If the model is the "brain," tools are its "hands & eyes"—letting it do things, not just say things.

Action tools: making AI agents do things

Some tools enable direct actions, allowing AI agents to interact with external systems and automate workflows. Example:

- Without tools: An AI can describe how to send an email.
- With tools: The AI sends an actual email through an API call.

Common action tools:

- Email API (Gmail, Outlook): AI can draft and send emails.
- Calendar API (Google, Outlook): AI can schedule meetings.
- Task Manager API (Trello, Asana, Notion): AI can create to-do lists.
- Finance APIs (Stripe, Plaid): AI can process transactions.

Retrieval tools: keeping AI grounded in facts

While action tools let AI agents "do" things, retrieval tools let them "know" things by fetching up-to-date or domain-specific knowledge. Example:

- Without retrieval: A customer service AI might give outdated refund policies.
- With retrieval: The AI fetches the latest refund policy before responding.

Common retrieval tools:

- Vector databases (Pinecone, Weaviate, Chroma): AI can store and retrieve knowledge.
- Enterprise search APIs: AI can search company documents.
- News APIs (NYT, Bloomberg): AI can access the latest news.

Computation tools: helping AI solve complex problems

Some AI agents need to perform calculations, run simulations, or analyze data beyond simple content generation.

Example:

- Without computation tools: AI can explain a physics formula.
- With computation tools: AI can run real calculations using WolframAlpha or Excel APIs.

Common computation tools:

- WolframAlpha API: Solves equations and computes results.
- Google Sheets API: Updates spreadsheets and processes data.
- Financial Modelling APIs: Runs pricing models, forecasts trends.

Why this matters for PMs:

Tool use expands what your product can do but it also introduces integration challenges, error-handling requirements, and decisions around what the agent should or should not be allowed to do. As a GenAI product manager, your job is to understand how these parts connect, what they enable, and what they constrain. You don't need to master every technical detail. But you do need to know enough to shape the product's boundaries, prioritize functionality, and ask the right questions like the following:

- Does this agent need memory?
- Should it be allowed to use tools?
- How will it recover if a step fails?
- What kind of planning will it need for the task it's being asked to do?

Agentic AI gives us powerful building blocks—but it's product thinking that turns them into real-world solutions. When you bring together a strong foundation model with orchestration, memory, tool use, and retrieval, something important happens. You move from abstract capability to practical execution. This combination forms the backbone of one of the most powerful agent architectures being used in real enterprise settings today—*Agentic RAG.*

Agentic RAG: Intelligent Retrieval Meets Action

Agentic RAG is a system architecture where agents don't just generate responses, they retrieve domain-specific knowledge, reason through tasks, and execute actions with precision. Think of it as a smart co-pilot that doesn't just consult the manual, it uses it to get the job done.

This is one of the most practical outcomes of combining the key components we've explored so far: foundation models, retrieval systems, tool use, memory, and orchestration. In enterprise settings, Agentic RAG enables AI agents to move beyond simple question-answering and into coordinated, goal-driven task execution. Here's how it works (see Figure 3.3 for a visual representation):

1. The user query is received by the agent.
2. The agent consults an LLM for reasoning.
3. The agent determines if external retrieval or tool use is required.
4. If memory is enabled, the agent recalls prior interactions (short-term or long-term).
5. The agent plans a response using techniques like ReAct or CoT reasoning.
6. The agent returns a response after synthesizing retrieved data and model reasoning.

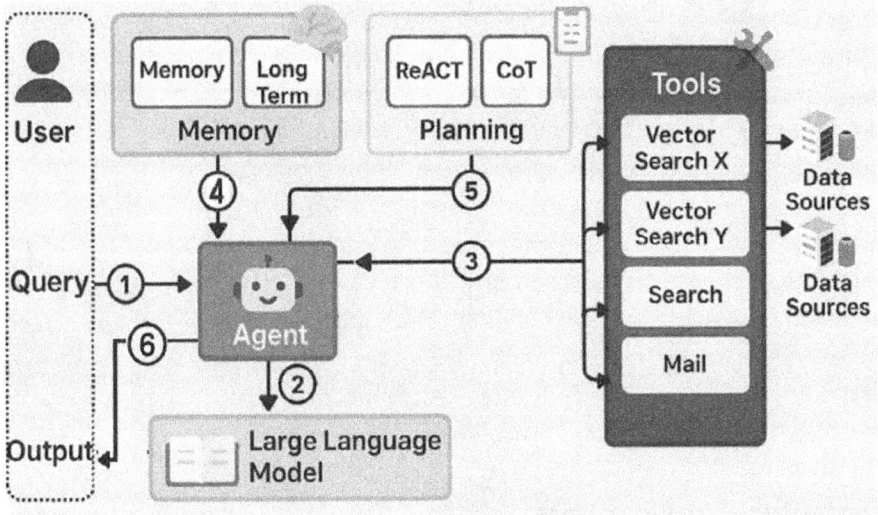

Figure 3.3: A visual representation of an agentic RAG system
© 2025 by the author of *Zero to GenAI Product Leader*.

As shown in Figure 3.3, Agentic RAG enables AI systems to move beyond simple content generation by retrieving domain-specific knowledge, reasoning through tasks, and executing actions. It's one of the most practical and widely adopted agent architectures in use today, especially in enterprise settings where coordination and factual accuracy are key.

But Agentic RAG is not a standalone category. It's best understood as an implementation pattern that sits somewhere along the broader spectrum of agentic AI maturity. Depending on how it's designed, how much planning, memory, autonomy, and coordination it supports—it may represent a moderately sophisticated system or something far more advanced.

To better understand where architectures like Agentic RAG fit in and to evaluate what level of agentic intelligence your product needs, it helps to look at the broader landscape. The next section walks through five foundational levels of agentic AI, each representing a different balance of capability, control, and complexity.

Levels of Agentic AI Systems

So, what does this spectrum of autonomy look like in practice? Agentic systems can operate at very different levels of sophistication—ranging from AI that simply replies to prompts, to systems that dynamically use tools, to full-fledged multi-agent networks capable of coordinating across tasks and domains.

These aren't just technical distinctions, they shape product scope, system reliability, user expectations, and development complexity. As a product manager, understanding where your solution falls on this ladder helps you make smarter decisions about architecture, feature design, and rollout strategy.

To make sense of this range, we'll use a five-level framework a practical way to classify agentic systems based on how much reasoning, memory, planning, and autonomy they support. As of mid-2025, this model reflects how real-world agent systems are evolving and how companies are beginning to adopt them in production. Let's walk through each of these five foundational levels, each representing a different balance of control, capability, and autonomy.

Level 1: reactive agents

Reactive agents represent the simplest form of autonomy. These systems generate responses based purely on a single input prompt, with no memory, no awareness of prior interactions, and no ability to take action. They are stateless and operate in a fixed input-output loop.

For instance, a customer support chatbot that answers one-off questions without remembering anything from earlier in the session fits squarely in this category. These agents are suitable for well-bounded use cases like FAQs or text generation, but they offer no continuity, adaptability, or depth.

Level 2: context-aware agents

At this level, agents begin to show continuity within a session. They can remember earlier turns in a conversation and adjust their responses based on that short-term memory. While they don't retain memory across sessions, they do provide a more coherent user experience in multi-turn dialogues.

An example would be an AI writing assistant that remembers earlier paragraphs in the current draft or a chatbot that refers back to something the user mentioned five messages ago. These systems still lack true initiative or planning, but they begin to simulate intelligent flow within a bounded context window.

Level 3: tool-using planners

Tool-using agents extend beyond conversation. They not only generate responses but can decide to invoke external tools—like APIs, calculators, or databases—to complete tasks. They often employ structured reasoning strategies like Chain-of-Thought or ReAct, allowing them to decompose tasks, gather supporting information, and generate more grounded and actionable outputs.

Consider an AI travel assistant that, given a goal like "book me a flight," breaks that down into subtasks, queries a flight API, compares options, and returns a recommendation. These agents blur the line between chatbots and workflow engines, offering value in areas like e-commerce, research, and productivity.

Architectures like Agentic RAG often operate at this level when agents combine retrieval techniques with tool APIs and structured reasoning to generate grounded, context-aware responses. While capable, these systems are still largely reactive—they act in response to a user's prompt rather than initiating action independently.

Level 4: autonomous agents

Autonomous agents take a significant leap forward: they are goal-driven systems that initiate actions on their own, adapt over time, and respond to dynamic outcomes in the environment. Unlike Level 3 systems, which rely on prompt-response cycles, autonomous agents are proactive—they can execute multi-step workflows based on a high-level objective without requiring human intervention at each step.

Picture an internal enterprise agent responsible for monitoring product feedback. Rather than waiting for a user to ask, it continuously tracks incoming reports, clusters similar issues, generates summaries, and alerts the relevant team when thresholds are crossed. These agents require persistent memory, access to a wide range of tools, and feedback loops that support learning and correction over time.

Agentic RAG architectures can evolve into this level when they incorporate goal-setting, persistent memory, and autonomous reasoning effectively shifting from reactive orchestration to proactive execution. This transition marks the onset of true autonomy, but it also brings new challenges around control, consistency, and system reliability.

Level 5: multi-agent systems

At the frontier of current capabilities are multi-agent systems—collections of specialized agents that collaborate to solve complex tasks. Typically, one agent acts as a coordinator, assigning subtasks to other agents based on their roles, context, and skills. These systems mirror organizational structures, with agents acting like teams: researching, synthesizing, and iterating together.

For instance, an AI-led research pipeline might have one agent finding sources, another summarizing them, and a third organizing the results into a final deliverable, all overseen by a managing agent. These systems unlock scalability and parallelism but also introduce complexity in monitoring, governance, and debugging.

Applying the framework

For most PMs, Levels 1 through 3 represent the current product-ready sweet spot. They allow for targeted automation, deliver user value, and integrate with existing systems without requiring radical infrastructure changes. Levels 4 and 5 are emerging fast, especially in enterprise and developer environments—but they demand a shift in how teams think about design, risk, and orchestration. Understanding which level your product needs and why will help you scope your roadmap, plan for guardrails, and avoid over-engineering.

The Evolution of AI: From LLMs to Autonomous Systems

GenAI has evolved from basic models to advanced, autonomous systems, reflecting a journey toward greater intelligence and integration. This progression spans four key dimensions:

- Knowledge Access (how the AI accesses information),
- Autonomy Level (how independently it operates),

- Integration (how it connects with external systems),
- and Key Capabilities (what it can do).

Let's explore this journey, building on the foundations we've covered, including the building blocks of agentic AI we discussed earlier. Figure 3.4 below maps this progression across eight approaches, showing how each step forward enhances the AI's abilities:

Approach	Knowledge Access	Autonomy Level	Integration	Key Capabilities
LLMs	Static from training	Low	Simple	Text generation, pattern recognition, basic reasoning
RAG	External retrieval	Low	Medium	Contextual responses with external knowledge
Tool/Function Calling	Connected services	Medium	Medium	API integration, system interaction
AI Agents	Dynamic via tools	Medium-High	Medium-High	Goal-oriented planning, tool selection
Agentic RAG	Intelligent retrieval	High	High	Self-directed information retrieval & synthesis
Graph RAG	Knowledge graphs	High	High	Relational understanding, causal reasoning
Multi-Agent Systems	Shared & distributed	Very High	Very High	Collaborative problem-solving, specialization
Model Context Protocol (MCP)	Universal knowledge	Very High	Very High	Cross-system standardization, semantic interoperability

Figure 3.4: AI Evolution: from LLMs to multi-agent systems to MCP—How GenAI systems have progressed across knowledge access, autonomy, integration, and capabilities.

Early stages: LLMs to RAG

The journey begins with traditional LLMs, which rely on static training data and simple integration to generate text, like drafting a story, though they lack autonomy. RAG systems take a step forward by accessing external knowledge, enabling grounded responses like answering a customer query from a company FAQ with medium integration but still limited autonomy.

Intermediate stages: tool use to agents

Tool/function calling marks a middle stage, connecting to services via APIs with medium autonomy and integration, allowing tasks like fetching live data from a search engine. The evolution continues with AI agents, which show increased autonomy through dynamic tool usage and goal-oriented planning, perfect for tasks like scheduling meetings, with stronger integration to external systems.

Advanced stages: agentic systems to MCP

More advanced systems include agentic RAG and GraphRAG, both with high autonomy and integration. Agentic RAG retrieves, reasons, and acts, ideal for multi-step tasks like contract analysis, while GraphRAG adds relational and causal reasoning, shining in research by analyzing connections, such as linking customer feedback to features. Near the apex are multi-agent systems, leveraging shared knowledge with very high autonomy and integration, enabling collaborative problem-solving through specialized sub-agents, like a team researching market trends. At the pinnacle is the Model Context Protocol (MCP), the most advanced approach, offering universal knowledge access, very high autonomy and integration, and cross-system standardization allowing AI to seamlessly sync data across platforms like calendars and project tools.

This evolution shows how GenAI systems have gained sophisticated knowledge access, greater autonomy, enhanced integration, and complex abilities moving from basic generation to semantic interoperability. But before we move forward, let's consider another foundational aspect: how these advanced systems are used by people and developers in practice.

How Users and Developers Consume GenAI Products

As GenAI systems evolve, they can be used in different ways, depending on who they're designed for, this is called the consumption method. Understanding these methods helps you think about how your product will be experienced, which will be crucial when you start building it in Chapter 5. There are two main ways GenAI products are consumed: through user interfaces (UI-based) and code-first methods.

UI-based products are designed for end users, offering a graphical interface that's easy to interact with. For example, a chatbot might let users type questions in a web app, or an AI assistant might schedule meetings through a mobile app with buttons and notifications. These interfaces are intuitive for users who don't need technical skills to use the AI.

Code-first products are designed for developers, offering programmatic access to the AI's capabilities. This can happen through different interfaces:

- *APIs (application programming interfaces)* allow developers to query the AI from their own applications—for example, an API might let a developer ask the AI a question and get a response to use in their app.
- *CLIs (command-line interfaces)* let developers interact with the AI using text commands—for example, a command might tell the AI to schedule a meeting.
- *SDKs (software development kits)* provide tools for developers to build AI features into their software—for example, an SDK might let a developer add AI suggestions to a coding tool.

For example, a GenAI product might offer a chatbot UI for end users to ask questions, but also an API for developers to integrate the same chatbot into a third-party app. As a GenAI product manager, you don't need to code or build these interfaces yourself, but you should be aware of these consumption methods. They'll influence how you design and build your product, ensuring it meets the needs of your target audience whether that's end-users, developers, or both. With this foundation in mind, let's wrap up our exploration of GenAI basics and look ahead to what's next.

Moving Forward

In this chapter, we've explored the foundations of generative AI, how it works, why it matters, and what's emerging with agentic AI, including how these systems are consumed by users and developers. But understanding technology is just the beginning. In the next chapter, we'll shift focus: What exactly does a GenAI product manager do in this evolving landscape? What makes the role different and how can you prepare for it?

Chapter 4: The Role and Skills of the GenAI Product Manager

In the GenAI era, product managers aren't just defining what to build, they're becoming co-builders of the future.

In March 2025, a Wuhan startup called Monica launched Manus—an autonomous AI agent that roams the web and handles tasks solo, from research to bookings, no human hand-holding needed. It's a glimpse into where we're headed: high stakes, fast pivots, and PMs steering systems that act, not just react.

I've felt this shift firsthand. Not long ago, AI was a quiet assistant in background helping users, not leading workflows. Today, it's the engine driving actions, collaborating with humans, and in some cases, operating independently. As AI evolves from a passive tool to an active teammate, the role of the product manager must evolve, too.

In this chapter, we'll explore the new mindset, skills, and responsibilities shaping the modern GenAI PM and why thinking across the full AI stack, not just at the feature level, is now essential for success.

Let's take a look at how this is already playing out in the real world.

One example is Microsoft 365's new generation of agents, rolled out in late 2024. These autonomous copilots flag critical emails, book meetings, surface insights, and drive workflows—all without requiring explicit, step-by-step instructions from the user. They're not just automating simple tasks, they're anticipating needs, making decisions, and keeping work in motion. That's agentic AI in action: systems that plan, remember, and execute. Systems that go beyond predicting the next best word and start orchestrating meaningful actions across apps and services. We're seeing this pattern repeat across the industry:

- GitHub Copilot evolving from code suggestions to multi-step coding agents.
- Notion AI helping users create full project plans or knowledge hubs based on loose prompts.
- Expedia's AI concierge adjusting entire travel itineraries on the fly based on real-time events.

These aren't static tools anymore; they are adaptive partners. And they represent the future that product managers must now design for.

As product builders, we're not just witnessing a shift we're part of it. Customers aren't simply using tools anymore; they're collaborating with intelligent partners, often without even realizing it. And the systems we design will define how natural—and how trusted—that collaboration feels. Chapter 2 explored how AI evolved from passive assistant to active collaborator. Chapter 3 laid the technical foundations behind this shift—memory, planning, tool use. Now, in Chapter 4, the focus turns to *us*, the product managers responsible for shaping these dynamic systems into trustworthy, intuitive, high-impact experiences.

Because as AI starts taking action, not just providing output, the stakes—and the expectations for PMs—rise sharply.

Some of the most effective GenAI product leaders are already adapting to this shift by rethinking how they learn, build, and lead. ***Amit Ghorawat (Director of Product Management at Reddit)*** points to the power of curiosity and peer learning—encouraging teams to share GenAI-driven wins and even skip lengthy PRDs in favor of rapid prototypes. ***Shambhavi Rao (Gen AI Product Lead at Google Search)*** highlights the importance of trend fluency—carving out time to explore broader tech shifts, even outside her product area, to anticipate where the puck will go next. And ***Selena Zhang (AI Product Leader at a global social media company, formerly at Amazon)*** emphasizes balancing fearless experimentation with domain safety, ensuring every feature is grounded in real-world trust constraints.

These aren't abstract philosophies. They're proof points from the field—showing how great PMs build systems that work and earn confidence as their tools begin to think for themselves. Their advice echoes a larger pattern we'll explore next: the mindset, skills, and behaviors required to thrive as a GenAI PM.

Setting the New Expectations for GenAI PMs

As AI systems become active decision-makers, you'll face new expectations as a GenAI PM. It's no longer enough to design predictable workflows, you need to create systems that adapt and maintain user trust, even when things get unpredictable.

This demands a reoriented mindset:

- **Design for behavior, not just features:** Your product's success hinges on how it acts, not just what it does. That means prioritizing transparent interactions—like making sure your agent's reasoning is clear—so users can trust its decisions.
- **Embrace variability and plan for ambiguity**: Non-linear outputs from models and agents are part of the deal, so you'll need to design recovery options that let users stay in control when results don't go as expected.
- **Build trust through collaboration**: Reliability can falter at the system's seams—think stale data retrieval, latency delays, or unclear interfaces. You'll need to work closely with teams across models, infrastructure, and user experience to prevent these failures and foster confidence over time.

This hands-on approach means orchestrating solutions across the AI stack to deliver seamless outcomes. As a GenAI PM, you're not just launching features, you're enabling intelligent systems that evolve with every interaction, setting the stage for the future of work. This mindset, embodied by the full-stack product maker, shapes the approach we'll explore next.

Note: As you read ahead, you'll notice I sometimes use "product manager" and sometimes "full-stack product maker." In the GenAI context, they are one and the same. Whenever you see "product manager," think of it as stepping into the expanded role of a full-stack product maker, someone who builds, orchestrates, and stewards intelligent systems end-to-end.

From Product Managers to Full-Stack Product Makers

The GenAI era doesn't just nudge product managers to adapt, it demands a reinvention. No longer are you the strategist sketching roadmaps from the sidelines. You're in the trenches, building alongside engineers, researchers, and designers, stitching together systems that think, act, and evolve.

This is the full-stack product maker: a product manager who doesn't just define what to build, but helps shape how the system behaves across every layer of the AI stack.

I've seen this shift up close. A year ago, a PM might have handed off specs and waited for the magic to happen. Today, you're prototyping workflows, tweaking prompts, or orchestrating agents to solve real user problems. It's high stakes and fast-paced, think of Monica's Manus, autonomously booking flights, or Notion

AI spinning up project plans from a single prompt. These aren't static tools; they're dynamic systems that demand a hands-on mindset. As a full-stack product maker, you're not just defining what to build, you're shaping how it behaves, across every layer of the AI stack.

Why Full-Stack Matters

Traditional PMs could lean on specialization, trusting engineers to handle the tech and designers to nail the UI. GenAI upends that. A chatbot's failure isn't just a bad prompt, it could be stale data retrieval, a latency spike, or an interface that confuses users. These systems fail at the seams, and you're the one who has to see the connections. Full-stack product makers don't need to code like engineers or train models like researchers, but they must understand how models, infrastructure, and user experiences intertwine to deliver trust and impact.

We saw a striking example of this in 2024, when Air Canada's generative AI–powered virtual assistant gave a customer incorrect information about ticket refunds. The court-ordered compensation was relatively minor—about CAD 1,400—but the reputational fallout was far more damaging. News of the mishap spread quickly, drawing attention to deeper issues across the stack: from retrieval safety to model oversight to poor fallback design.

This wasn't a model glitch, it was a system-level breakdown. And it illustrates why full-stack thinking isn't optional in GenAI product management. That systems view comes to life when you remember how *Saty Das, Staff PM at TikTok Shop*, approaches every project with a systems-thinking mindset understanding not just how the models work, but how people interact with them, and leaning into user context to design for trust at every layer.

In this space, small misfires don't just produce bad answers, they can erode user trust. And that's the real cost PMs must work to avoid. In GenAI, it's not the size of the initial mistake that defines the damage, it's the erosion of user trust. And that's why full-stack product makers are critical: to design resilient systems that prevent small misfires from becoming major failures.

In GenAI systems, no single layer can guarantee success alone. Even the best model can stumble without fresh retrieval. Even the cleanest UI can fail if latency spikes during critical workflows. And as we move deeper into the agentic AI era— where systems not only generate responses but initiate actions and plan workflows independently, this interconnected fragility only grows.

Great full-stack product makers connect these layers early catching weak seams, building resilience across orchestration flows, and designing for trust even when behavior becomes more dynamic and autonomous.

In this new landscape, where behavior is fluid and stakes are high, success demands more than static roadmaps—it demands a hands-on playbook *and* a full-stack mindset. Next, we'll explore both: how full-stack product makers *build differently*—and how they must *think differently*—to thrive in the GenAI era.

The New Playbook: Build, Don't Just Plan

Being a full-stack product maker means getting your hands dirty. You're in the mix—experimenting, iterating, and learning as you go, not waiting for someone else to prototype solutions or fine-tune models.

As Rocky Zhang (GenAI Product Leader, Airbnb) urges, "Don't wait to be assigned to an AI team to start learning. Try building something small—reverse engineer user experiences powered by LLMs and ask what's happening under the hood."

And, as *Saty Das (Staff PM, TikTok Shop)* reminds us, "You don't need the 'AI PM' title to dive in: Start learning by doing—build a prototype, experiment with APIs, or analyze how GenAI is being used (or misused) in the products around you. Curiosity compounds fast in this field."

This hands-on approach is organized around three pillars—Prototyping, Orchestrating, and Customizing—each reshaping how you deliver value in GenAI products.

Prototyping: test ideas early

Prototyping isn't just for engineers; it's your tool to validate ideas before they scale. With GenAI, you can spin up a proof-of-concept in hours, not weeks, using platforms like Azure AI Foundry and GitHub Copilot in agent mode to chain models, tools, and workflows without heavy engineering lift.

Consider Providence, one of the largest U.S. healthcare systems. In 2023, their team developed ProvARIA, a GenAI-powered assistant built with Azure OpenAI Service, designed to help triage and classify patient messages. In just 18 days, they created a working prototype and piloted it across four clinics—validating the concept with real medical assistants and providers before scaling further.

The pilot surfaced important gaps in how urgent messages were prioritized, which they refined by iterating on prompt design and routing logic, saving hours of manual triage work without heavy upfront investment. Prototyping builds credibility with technical teams, too. When you show up with a working prototype, engineers see you as a partner, not a bystander. It's not about perfection, it's about proving what's possible and learning where the system bends or breaks.

Orchestrating: choreograph intelligent systems

GenAI systems aren't solo performers, they're ensembles. As a full-stack PM, you orchestrate agents, models, and tools to work in harmony, like a conductor aligning an orchestra. Sometimes that means designing complex flows across multiple specialized agents. Other times, it means ensuring that different layers inside a single system—retrievers, reasoning models, and conversational interfaces—cooperate seamlessly.

Your job is to design the flow: deciding when a system should search versus generate, how it should recover if retrieval fails or returns ambiguity, and what guardrails ensure the experience stays trustworthy under real-world conditions. These decisions shape how intelligent the system feels and how resilient it becomes when things go off-script.

A standout example of this kind of orchestration comes from early 2025, when Physics Wallah (PW), one of India's largest EdTech platforms, introduced *Gyan Guru*—a GenAI-powered study companion designed to support millions of students seeking academic help. Rather than defaulting to a simple chatbot, PW took a layered approach. An agent first searched for a curated vector database containing three years of solved doubts and study material. Then, a GPT-based model—hosted via Azure OpenAI Service—generated natural-language explanations grounded in that retrieved content. A conversational interface managed multi-turn interactions, surfacing clarifications, retries, and follow-up queries as needed.

This flow—retrieval first, generation second, and dialogue throughout—transformed static content into a dynamic, always-available tutor. Crucially, the PW team didn't just prioritize speed—they optimized for trust. They constrained model creativity, validated retrieved content, and tested Gyan Guru under real-world student interactions before full deployment. The result wasn't just a helpful assistant—it was a choreographed system built for learning resilience. Gyan Guru demonstrates that orchestration is more than just technical design, it's a product decision at every step.

Full-stack PMs don't just connect AI components, they design invisible choreography that builds trust, resilience, and delight at scale. It's not about manually controlling every decision. It's about shaping intelligent systems that behave predictably, recover gracefully, and deliver trustworthy outcomes, even as underlying technologies evolve.

In agentic AI systems, orchestrating memory, reasoning, and dialogue layers is as critical as coordinating multiple agents across workflows. This layered approach ensures cohesive, trustworthy interactions, especially in complex domains like education.

Customizing: tailor for impact

The power of GenAI lies in its flexibility but off-the-shelf models and generic workflows rarely deliver lasting value. Full-stack PMs play a key role in shaping systems to fit real-world contexts, partnering with researchers and engineers to adapt AI for specific domains, tone, and trust requirements. This customization goes far beyond prompts. It includes tailoring output style, refining workflows, aligning with regulatory standards, and localizing for audience expectations.

For example, a product manager building for healthcare might adjust a chatbot's tone to express empathy with patients, or fine-tune retrieval systems to surface only regulatory-compliant content in financial services. The real skill lies in knowing *how much* to tweak—and *when* to intervene—whether it's rewriting a prompt, fine-tuning a model, or switching approaches altogether. These decisions define whether a GenAI product simply works or truly resonates.

One compelling example comes from Chi Mei Medical Center, one of Taiwan's largest hospitals. In 2025, they partnered with Microsoft to build a suite of specialized AI copilots on Azure OpenAI Service. Instead of deploying a one-size-fits-all chatbot, Chi Mei created a family of agents tuned for specific roles: pharmacy support, doctor report generation, and nurse shift coordination.

What made it work wasn't just the tech; it was the iterative customization process. Teams fine-tuned prompts, adapted language to match medical norms, corrected terminology, and embedded retrieval from internal databases to ensure clinical accuracy. They didn't just integrate the model into existing workflows—they shaped it until the outputs felt intuitive and trustworthy to frontline staff.

The results were measurable and meaningful: faster documentation, doubled throughput in pharmacy workflows, reduced clinician burnout, and high adoption across departments. This wasn't a story of "smarter models"—it was a success born from deliberate, product-driven customization. As ***Rocky Zhang, Gen AI Product Leader at Airbnb,*** puts it, evry technological shift is an invitation to build intuition from first principles. You don't need to be an ML expert, he says, but you must get comfortable making product decisions under uncertainty—and that mindset defines the difference between generic AI and truly helpful solutions.

Chi Mei's approach reinforces a core lesson: in GenAI, customization isn't a last-mile enhancement—it's the reason users trust and adopt your system in the first place.

The Mindset Shift: From Planner to Builder

Becoming a full-stack product maker isn't just about gaining new skills, it's about operating with a fundamentally different mindset.

In traditional product management, success meant planning features carefully, writing crisp specs, and managing predictable workflows. But GenAI systems don't behave predictably. They evolve. They drift. They interact with users and data in ways that no roadmap can fully capture.

In this new landscape, product managers can't just plan around uncertainty they must build through it. You're not sketching features from the sidelines anymore. You're prototyping workflows, tweaking prompts, orchestrating agents, stitching models, APIs, and UX flows together in real time.

Success now demands three core habits:

- **Embrace experimentation:** Launch rough prototypes early. Test assumptions fast. Let real-world feedback guide iteration, not internal perfectionism.
- **Own the end-to-end:** From the initial idea to production deployment, take accountability. Dive into orchestration flows, debug APIs, adjust model behavior—own the full stack experience, not just your slice of it.
- **Earn technical trust:** Show up with working demos, prototype flows, and tangible improvements. Partner with squads of engineers, researchers, and designers, speaking their language to collaborate as equals.

This isn't just a change in tactics—it's a whole new way of thinking. ***Sunny Tahilramani, Gen AI Product Leader at Google***, reminds us that the winners in this era think in platforms, not features, and empower others to build with them.

Equally important is a curiosity-first approach to your own growth. As **Mark Cramer (Sr PM, Meta)** puts it:

"I think more generally it's worth thinking about mindsets and skills to enable adaptation to any new technology. In my particular case it was Machine Learning broadly and the mindset I applied was one of curiosity and continuous learning.

Andrew Ng has a great quote about lifelong learning. The part I found particularly interesting was that lifelong learners should, 'seek knowledge or skill beyond what would be immediately useful.' This is what led me to pursue learning how to program (Java on Android and then Python), taking some online courses in statistics and machine learning, and then obtaining an AI graduate certificate from Stanford.

None of that was immediately useful and, in fact, was most likely overkill for a PM, even an ML PM. I'm confident, however, that an ML side project got me a job at PARC and graduate coursework really helped when I joined Meta's Video Core ML team. I did it, however, not because I saw an immediate (or any) career boost, but because I was fascinated by the technology and curious about how it works. I strongly believe that anyone and everyone can be a lifelong learner and follow their own path."

That same curiosity-first mindset, combined with platform thinking, is what lets full-stack product makers build dynamic, evolving systems—systems that behave intelligently, recover gracefully, and earn user trust even as models, infrastructure, and needs keep changing.

This mindset evolution from planner to builder, from surface to stack is what separates good GenAI PMs from great ones.

Why it matters:

The full-stack product maker role isn't just a job; it's a mindset that unlocks impact. You're not waiting for the system to catch up; you're making it happen, from the first prototype to the final rollout. You're earning trust not just from users, but from teams who see you as a partner in the trenches. And in a world where GenAI systems evolve with every interaction, you're not just keeping pace— you're setting the stage for the future of work.

This hands-on approach shapes the core responsibilities PMs undertake, from framing problems to connecting the stack, which we'll detail next.

Note: The evolution from traditional PMs to full-stack product makers reflects a broader industry trend I've observed through conversations and observations of GenAI product leaders—particularly around building, orchestrating, and leading intelligent, evolving systems.

Core Responsibilities

Understanding how generative and agentic AI work is just the beginning. As a GenAI PM, your job isn't to build the models, it's to make them reliable, intuitive, and trustworthy inside real-world products.

These responsibilities cut across every role in the stack—whether you're scaling infrastructure, fine-tuning models, orchestrating agents, or designing ethical guardrails. You're not just launching features; you're steering systems that learn, generate, and sometimes fail in unexpected ways. Even if your ownership is narrow—say, model tuning or deployment quotas—your awareness must be broad. A stale retrieval result can tank a good UI. A slow endpoint can break trust in a helpful response. A vague prompt can overwhelm a perfect model. Your success depends on connecting dots across the stack. The fundamentals stay the same: define value, manage risk, earn trust. But how you do it depends on where in the stack you operate—and how well you connect to the rest.

Let's break down those shared responsibilities and how they shape your work as a GenAI PM.

Understand Capabilities and Gaps

You start by knowing what AI can pull off and where it trips. Can it draft emails but choke on reasoning? Will an agent scale or crash? Get this wrong, and you're sunk before you begin. The key takeaway is this: ambition must match capability, test limits, and not just trust specs. Push your systems in real-world conditions before betting the product on them.

Framing the Right Problem for AI

AI isn't a one-size-fits-all solution. It thrives in some problem spaces and flops in others. As a GenAI PM, your first job is to ask the right question:

- Is this a problem generative AI is good at solving?
- Should the AI generate, retrieve, decide, or simply assist?
- Can the solution scale safely and ethically?

Frame it right, or it's just noise chasing tech, not impact. Take Manus in March 2025, it excels at booking flights but struggles with debating ethics. The lesson is clear: match the problem to what AI does best.

Designing Human-AI Collaboration

Great GenAI products don't just deliver answers, they shape workflows. As a full-stack PM, your job isn't just to optimize the model, it's to choreograph the handoff between human and machine. Where does the AI take the lead? When should it pause, ask for input, or hand back control?

This new interface paradigm requires more than clever UX, it demands a deep understanding of trust. Users need clarity about what the system is doing, why it's doing it, and how they can intervene. Can the AI explain its reasoning? Can users steer the outcome, provide corrections, or escalate uncertainty?

A powerful example of this design in action is GitHub Copilot. As developers type, the AI offers inline code suggestions in real time. But those suggestions are not final—they're editable, ignorable, or acceptably integrated into the developer's own work. The product doesn't just inject intelligence into the flow; it respects the flow. Copilot empowers users while supporting them, offering a fluid dance between human intent and AI augmentation.

That's the goal of human-AI collaboration design: to create experiences that feel intelligent but remain anchored in user control. When done right, the AI becomes not a replacement for the human, but an extension of their capability.

Track AI in Action

Shipping a GenAI feature isn't the end, it's the beginning of active oversight. Once your system is live, its performance becomes dynamic, shifting with new data, evolving user behavior, and edge-case surprises. As a full-stack PM, your role is to monitor not just whether the system works, but how well it works in the real world.

It's not just about the model's raw accuracy. You're responsible for how the entire system behaves, from latency and stability at the infrastructure layer, to the reasoning quality of model outputs, to the clarity, safety, and fairness of the user experience. Each layer might be owned by different teams, but users experience them all as one.

That's why successful PMs don't rely only on static metrics. They tune into how the product feels in motion, how it responds under stress, adapts to ambiguity, and recovers when something breaks. If the AI starts drifting, you need to catch it early. Because in GenAI, trust is fragile and what seems like a small misfire internally can quickly become a confidence-breaking flaw externally.

Ship Fast, Think Risk

GenAI systems don't just evolve through code updates, they evolve with data, context, and even prompt phrasing. A single change in model version or retrieval logic can alter behavior in ways you didn't anticipate. That's why product managers working on AI systems must internalize a new mantra: ship fast but think in risk.

You're not just building features—you're managing fluid, probabilistic behavior. That means testing, safety, and version control aren't checkboxes at the end. They're part of the product's DNA from day one. Successful teams monitor drift constantly, catching when models begin veering from expected behavior. They integrate compliance and safety checks early, not as afterthoughts. They use feature flags, canary rollouts, and shadow testing to de-risk launches.

In March 2025, Anthropic rolled out web search integration for Claude, including a critical feature: citations. This wasn't just a nice-to-have—it was a direct response to early failures where hallucinated answers eroded user trust. By anchoring outputs with links to sources, they didn't just boost transparency—they rebuilt confidence (Wiggers 2025).

The lesson? You can—and should—move fast. But in GenAI, speed without foresight can backfire. Guardrails aren't constraints; they're what make fast innovation sustainable.

Connecting the Stack End-to-End

Even if you only own one part of the stack, users experience everything. They don't care where the glitch came from—just that something felt off. That's why great GenAI PMs think horizontally, not just vertically. A biased model, stale retrieval results, a slow API call, or a clunky interface—any of these can break trust. Your job is to connect the dots:

- Model behavior: Is it accurate and fair?
- Retrieval: Is the info fresh and relevant?

- Tool use: Are actions timely and reliable?
- Infra: Can it scale under real-world load?
- UX: Does the user feel in control?

You're not just PM-ing your slice of the pie, you're stitching together an end-to-end experience that works.

These responsibilities may sound familiar but in GenAI, executing them demands sharper instincts and deeper empathy. You're not just managing features; you're orchestrating behavior, managing unpredictability, and earning trust across evolving systems.

Now that we've seen *what* full-stack product makers must own, let's dive into *how* they build the capabilities to lead in this dynamic GenAI world. First, we'll explore how responsibilities show up across different stack layers. Then, we'll break down the essential skills that help PMs succeed across this complexity.

Mastering the GenAI Stack: How PMs Lead Across Layers

We've just explored the responsibilities that unite all GenAI PMs—framing the right problems, shaping human-AI interaction, managing risk, and tracking system behavior. But how do those responsibilities show up in your day-to-day work? That depends on where you operate in the GenAI stack.

A PM focused on infrastructure thinks in terms of GPU provisioning and quota logic. A model PM balances fine-tuning, hallucination risk, and evaluation pipelines. A UI PM focuses on trust, human oversight, and explainability. Leading squads of researchers, engineers, and designers, they prioritize user needs, with agent experience (AX) growing prominent, as we'll explore later.

The fundamentals stay the same—but the expression of the role shifts layer by layer. While PMs may specialize in a specific layer, the best PMs master their layer while seeing its ripple effects across the stack. Seeing the whole picture is what unlocks real leadership. Figure 4.1 below provides a high-level view of how PM responsibilities evolve across each layer of the GenAI stack.

GenAI Stack Layer	Core PM Focus	Full-Stack Mindset Cue
Infrastructure	Scaling computer power, storage, networking	Slow systems break user trust
Model	Shaping helpful, safe, efficient AI models	Model choices shape trust downstream
AI Logic & Execution	Orchestrating steps, tools, and data retrieval	Seams are where systems fail
Hosting & Serving	Deploying fast, reliable access points	Delays or crashes erode trust
Observability & Governance	Tracking, auditing, ensuring ethical behavior	Trust is maintained, not just built
AI Application & Interface	Designing intuitive, trust-driven interfaces	Interfaces shape how AI feels, not just looks

Figure 4.1: PM responsibilities across the GenAI stack
© 2025 by the author of *Zero to GenAI Product Leader*.

Let's zoom into the six-layer GenAI stack and explore how PM responsibilities evolve across it.

AI Infrastructure-Layer PMs

Powering the foundation behind every AI experience, nothing happens without raw compute power. That's where infrastructure PMs come in. They work behind the scenes but without them, GenAI features don't launch, models don't scale, and inference falls apart.

They manage the plumbing—compute, storage, and networking—across cloud, edge, or on-prem systems. That includes making sure the right GPUs or TPUs are available, but also ensuring models have fast access to data and can communicate efficiently across services. It sounds technical and it is but it's also deeply product oriented. Their choices determine whether AI feels snappy or sluggish, affordable or impossible to scale. They work across teams—model PMs, research leads, DevOps—to make sure the underlying systems evolve as the AI stack does—think bigger models or real-time Copilot shifts. I've seen launches die when infra lags. And when things go wrong? It's the infra PMs who set the guardrails—quotas, autoscaling logic, fallback plans—so the system fails gracefully, not catastrophically.

Full-stack mindset cue
Even if infrastructure PMs never touch the user interface, their decisions shape user experience profoundly. A delayed API call, a memory timeout, or a burst of latency from a saturated cluster can shatter trust in an otherwise well-designed product.

For example, a PM on Azure's AI platform might be responsible for GPU quota allocation when hosting Meta's Llama 3. Their work ensures that model traffic can scale dynamically during high-demand periods—without overloading cluster capacity or degrading performance. It's not flashy, but it's foundational. Invisible infrastructure, visible impact.

Why this matters:

Every second of delay, every unexpected crash, every spike in cost—it all starts (or gets solved) here. AI Infrastructure-layer PMs don't just support products; they enable the entire GenAI ecosystem to function at scale.

AI Intelligence-Layer PMs

This is where the AI magic begins—PMs here work on the very models that generate, predict, summarize, classify, or reason. But contrary to popular belief, most AI intelligence-layer PMs aren't training giant models from scratch—they're shaping how those models get used, evaluated, and improved for real-world products.

At this layer, you might be working with foundation models (like GPT, Claude, or Gemini), open-source models (like Llama or Mistral), or proprietary models built in-house. Your job isn't just to ask, *"Does the model work?"*—it's to ask, *"Is it helpful, safe, and worth the cost to run?"*

You own trade-offs—size vs. latency, quality vs. compute cost, general vs. fine-tuned. You define what "good" looks like: accuracy, helpfulness, tone, bias mitigation, or even carbon footprint. And you obsess over failure modes—hallucinations, overconfidence, or refusal to answer—because these aren't bugs, they're user-facing behaviors. You're also the bridge between AI researchers, infra teams, and application PMs helping everyone understand what a model can (and can't) do, when to fine-tune, when to RAG, or when to rewrite the prompt instead. Model PMs often create internal tooling too—evaluation dashboards, telemetry pipelines, safety classifiers—that help other teams deploy models responsibly. If AI infrastructure-layer PMs make AI possible, AI intelligence-layer PMs make it usable.

Full-stack mindset cue
Model choices ripple downstream. Whether you're adjusting prompts, fine-tuning outputs, or balancing cost and quality—those decisions shape what infra must scale, what UX must explain, and how trust is earned. You're not just optimizing models—you're setting the tone for the entire system.

Models are the face of the product. If they hallucinate, lag, or can't adapt, the whole system breaks trust. PMs here shape how smart, useful, and safe GenAI actually feels to end users.

AI Logic & Execution-Layer PMs

This is where models stop being static and start behaving like intelligent systems. PMs in this layer focus on how AI "thinks," plans, remembers, and acts. They don't just ship prompts or APIs, they choreograph how AI flows across tasks, tools, and user intent.

You're designing logic: sequencing steps, deciding when to use memory, when to call tools, how to resolve ambiguity, and how to handle edge cases. This could involve building with orchestration frameworks like LangChain, Semantic Kernel, or AutoGen, or managing custom pipelines that power agentic behavior. You decide what happens when the system gets stuck, how long tasks are coordinated, and when to escalate to a human.

You also own how models connect to the outside world—retrieval systems, tools, APIs, and data sources. When the AI responds, did it ground its answer in trusted knowledge? Did it pull the right data before taking action? That's on you. And increasingly, you may be stitching together multiple models—routing queries across them based on context, cost, or task fit. Whether it's GPT-4 for reasoning, Claude for summarization, or a domain-specific model for compliance— you decide how the system chooses and chains them. But orchestration isn't just technical glue— it's product logic. What sequence of decisions leads to user success? Where should the system pause? What if it breaks mid-flow? Your orchestration defines not just what the AI can do— but how it behaves.

Full-stack mindset cue
This layer is where seams are most visible. A shaky retrieval system can break trust. A bad tool call can confuse users. Your orchestration logic must align with model behavior, infra reliability, and UX guardrails—because you're the connective tissue.

Why this matters:

Great models aren't enough. Without orchestration, GenAI stays static—just prompt in, output out. PMs in this layer turn isolated model calls into coherent, intelligent behavior. They bring agents to life.

AI Hosting & Serving-Layer PMs

The smartest model means nothing if it can't respond fast and reliably at scale. AI Hosting and serving-layer PMs make GenAI usable in the real world ensuring models and agents can be deployed, accessed, and trusted across environments. This layer is responsible for packaging, deployment, scaling, and inferencing. You define how models and agentic workflows are exposed: via APIs, serverless endpoints, or containerized workloads—whether they run in the cloud, at the edge, or on-prem.

You'll work closely with AI infrastructure-layer PMs, who provide the raw compute (like GPU clusters or TPUs). Your job is to build productized interfaces on top—so that developers don't worry about capacity, latency spikes, or cold starts. You also own inference performance. That means selecting or supporting optimized runtimes (like ONNX Runtime or TensorRT), reducing model load time, tuning for concurrency, and driving down cost per request. It's a mix of performance engineering and product experience.

As agentic use cases grow, you'll also define how agents are hosted—supporting persistent memory, tool calling, and long-running workflows across multiple steps. Hosting now isn't just about outputting text—it's about supporting stateful, multi-step intelligence. And when things break—timeouts, errors, unresponsive endpoints—you manage service-level agreements (SLAs), retries, observability hooks, and escalation paths to keep things running smoothly.

Full-stack mindset cue
A sluggish model or unstable endpoint undermines everything upstream. You may not own the model logic or UX, but your choices shape how intelligence feels: fast, helpful—or broken and frustrating.

Why this matters:

GenAI systems aren't just about what the AI knows, they're about whether it shows up on time and performs under pressure. Hosting PMs turn potential into production.

AI Governance, Observability & Responsible AI-Layer PMs

You can't fix what you can't see. And with GenAI, that visibility needs to go far beyond uptime or usage stats. PMs in this layer build the systems that monitor, debug, and govern how AI behaves in the wild—because trust isn't just built during development; it's maintained after deployment. Your work centers around making model behavior visible, explainable, and auditable. That includes building telemetry pipelines, dashboards, alerting systems, and evaluation frameworks that track things like the following:

- Drift in model behavior
- Hallucination rates
- Latency spikes or throughput bottlenecks
- User override patterns
- Safety violations or biased outputs

But observability isn't just technical—it's also about accountability and compliance. You'll work with legal, privacy, and ethics teams to define what responsible AI looks like in practice: logging decisions, tracking data lineage, surfacing explanations, and supporting human-in-the-loop workflows when needed. As AI becomes more autonomous, governance becomes product-critical. That means designing systems for the following:

- Auditing AI decisions
- Detecting and mitigating harmful or unfair outputs
- Enabling opt-outs, user control, and explainability
- Meeting regulatory requirements like the EU AI Act or industry-specific guardrails

Note: In many orgs, PMs focused on AI ethics and responsible AI initiatives also operate in this layer. While observability and governance may be separate in mature setups, we've combined them here to reflect how they often blend in practice.

Full-stack mindset cue
This is where trust gets real. A silent failure or biased output won't show up in logs unless you build for it. Great PMs here anticipate risk, connect across layers, and design guardrails into the core product—not just bolt them on after launch.

Why this matters:

In a world where AI drives decisions, observability isn't just operational—it's ethical. PMs here don't just protect systems—they protect users, organizations, and trust.

AI Application & Interface PMs

This is where users meet the machine. And in GenAI, that handshake has never mattered more.

PMs at the AI Application & Interface layer don't just build screens, they design trust. You're translating complex model behavior into interfaces that feel intuitive, helpful, and safe. Whether it's a prompt box, a chat interface, an AI-powered dashboard, or a copilot overlay, the goal is the same: make the AI feel like a partner, not a puzzle. You decide how users ask for help, see what the AI does, and course-correct when it slips. That includes the following:

- Designing user input patterns (freeform text, structured prompts, tool selection)
- Showing model reasoning or confidence levels (when appropriate)
- Enabling feedback, edits, and override actions
- Crafting escalation paths—what happens when the AI fails or is unsure?
- Avoiding "magic" moments that confuse rather than delight

You'll also partner deeply with design, research, and trust teams to shape transparency, explainability, and consent. GenAI experiences often feel magical but if users don't understand what happened or can't steer it, trust erodes fast.

Full-stack mindset cue
The interface isn't just cosmetic, it's behavioral. You're not designing a feature. You're designing how people collaborate with AI. Even small tweaks—like labelling a suggestion as "AI-generated"—can shape perception, confidence, and control.

Why this matters:

Even the smartest model fails if users don't trust or understand it. PMs here make GenAI feel human and that's what drives adoption, loyalty, and long-term impact. So far, we've looked at GenAI PM roles across the core stack—from infrastructure to interface. But some of the most critical product work doesn't sit *inside* the system, it lives *around* it.

That's where ecosystem PMs come in.

They focus on the connective tissue that scales GenAI to the world: onboarding third-party models, defining marketplace experiences, building monetization strategies, and growing developer platforms. They don't just think in terms of APIs or

UIs, they think in terms of networks, incentives, and trust at scale. Let's look at how their work shapes the broader GenAI platform economy.

Ecosystem PMs

Ecosystem PMs build the connective tissue between your platform and the world.

These PMs don't work *inside* the AI system—they scale it outward. EcosystemPMs focus on enabling others to build on top of your platform. That includes defining how third-party models, tools, and apps are discovered, integrated, trusted, and monetized. You might be working on a model catalog (like Azure AI Model Catalog or Google's Model Garden), building APIs for partners, or crafting the business and policy frameworks that govern how developers list, price, and earn from their GenAI assets.

This role blends platform thinking with business acumen. You're designing for *two* audiences: external contributors (model providers, startups, devs) and the end customers they serve. A misstep in monetization design, partner support, or customer trust can stall the entire ecosystem. You'll navigate partner onboarding, usage tracking, metering, performance standards, marketplace visibility, and revenue share models. Often, you'll collaborate with legal, marketing, engineering, and business planning to turn your ecosystem into a sustainable growth engine.

Full-stack mindset cue
Even if you don't own model behavior or infra scaling, your monetization and governance choices shape the entire system—who joins, what they build, how they're discovered, and whether the economics work.

Why this matters:

Ecosystem PMs stretch GenAI past one product. They link creators, users, and platform power—making it thrive, not just work. Without strong monetization strategies, discoverability flows, and partner tooling, even the best AI models can sit unused. PMs here turn platform value into real-world adoption and revenue.

One Role, Many Expressions

Across the six layers—and into the ecosystem—GenAI PMs show up in different ways. Some tune models. Some orchestrate logic. Some ensure tools run at scale. Others shape how the world discovers, trusts, and pays for AI.

But seeing the stack isn't enough. To lead in this space, GenAI PMs need more than feature management—they need sharp instincts, cross-layer technical fluency, and collaboration skills that bring dynamic systems to life. Let's explore the core skills that set full-stack product makers apart.

Core Skills of Full-Stack Product Makers

You've seen how GenAI PMs lead across the stack, now let's break down the core skills that make that leadership real. From navigating technical seams to shaping human-AI collaboration, these hard and soft skills turn full-stack thinking into products that users trust and rely on.

Hard Skills: Navigating Technical Complexity

As a full-stack product maker, you won't need to code the models or architect the servers, but you do need enough technical fluency to spot risks early, guide smart decisions, and keep evolving systems resilient.

One of your core responsibilities is understanding *model behavior and evaluation*. While you won't be training models yourself, you must grasp their quirks—like how prompt engineering influences outputs, when fine-tuning matters, and what metrics (accuracy, helpfulness, toxicity) reveal about model performance. This insight helps you anticipate failure modes, define limits, and test reliably before going live.

You'll also need to understand *tooling and orchestration*. GenAI systems rarely work in isolation—they rely on retrieval components, API chains, memory modules, and external tools to function effectively. Whether your team uses Semantic Kernel, LangChain, or a custom stack, your job is to ensure that these flows perform intelligently and consistently under real-world pressure—not just in sandbox demos.

Equally important is being aware of *data flow and latency*. Every user query travels through a complex pipeline, touching databases, networks, and models along the way. You need to be able to trace these journeys, identify latency bottlenecks, and work with engineers to optimize speed and responsiveness without sacrificing accuracy or reliability.

Finally, embrace *AI metrics and telemetry* beyond standard usage stats. Track override rates, grounding failures, model drift, prompt success, and user recovery behavior. These signals don't just inform product performance, they shape future iterations, help you diagnose issues early, and provide the evidence base for your roadmap decisions.

Soft Skills: Leading Through Ambiguity

Technical instincts can carry a product manager through specs and sprints—but in the GenAI era, soft skills become your stabilizers amid rapid change, emerging capabilities, and evolving user expectations. Your job is no longer just shipping features, it's guiding teams through complexity, aligning systems with user needs, and earning trust at every layer.

It starts with *problem framing in the AI context.* Not every problem needs AI, and not every AI solution adds value. One of your most critical skills is learning to frame problems that actually benefit from generative or agentic solutions—avoiding the trap of tech-for-tech's-sake. You'll need to calibrate ambition with feasibility, matching the strengths of today's models to real user pain points.

You'll also need to lean into *cross-functional collaboration.* GenAI product development is never a solo effort, it's a squad game. Researchers, infra engineers, prompt specialists, designers, data scientists, legal teams—each brings a unique lens. Your job is to synthesize these perspectives, set a clear direction, and move fast without breaking alignment. Whether it's translating model behavior into UX implications or aligning infra constraints with user promises, you lead by creating clarity across moving parts.

Another essential trait is *ecosystem and market awareness.* The GenAI landscape shifts weekly new models, API launches, open-source breakthroughs, emerging safety norms. You need to stay plugged into these trends, not just for curiosity's sake, but to ensure your product remains competitive, trusted, and timely. What worked three months ago may already be obsolete.

Finally, you'll need to develop what's increasingly known as an *Agent Experience (AX) mindset.* This goes beyond designing screens or interfaces, you're designing AI behaviors. That means thinking about recovery flows, user control, escalation paths, explainability, and transparency. Can the user tell what the AI is doing? Can they intervene when needed? Can they trust it to act on their behalf? These questions will define how safe and empowering your experience feels not just how well it performs technically.

Together, these soft skills form the foundation of trust-centered AI product leadership. They don't just help you ship—they help you lead when everything's shifting.

You've already seen how full-stack product makers think, build, and lead differently. Before we move ahead, let's crystallize the shift: how exactly does GenAI product management diverge from traditional product management expectations—and why does it demand a fundamentally different mindset?

How GenAI PMs Differ from Traditional PMs

GenAI PMs lead squads to shape adaptive, trustworthy systems unlike traditional PMs, who manage predictable, linear workflows. While traditional PMs ship features with clear specs and stable milestones, GenAI PMs navigate evolving systems where behaviors shift, and trust must be constantly earned. Across the stack—from powering infrastructure to designing human-AI collaboration—GenAI PMs guide behavior, not just deliver functions. Roadmaps become living hypotheses, refined continuously through feedback and system behavior. Figure 4.2 contrasts these roles, highlighting how the full-stack mindset demands deeper collaboration, sharper trust calibration, and faster adaptability.

Aspect	Traditional PMs	GenAI PMs
Core Goal	Ensure intuitive usability	Build user trust and resilience
Core Activity	Develop product features	Guide AI system behavior
System Behavior	Fixed outcomes, stable systems	Variable outputs, evolving behavior
Execution Model	Team-led execution	Co-creation with AI and squads
Approach to Errors	Error-free mindset	Failure-tolerant mindset (with fallback and escalation)
Testing Methodology	A/B testing on stable features	Rapid iteration with behavioral and trust tracking
Development Process	Validate and ship	Monitor, adapt, and evolve live systems
Planning Approach	Static roadmaps and milestones	Hypothesis-led iteration and dynamic planning

Figure 4.2: Traditional PMs vs. GenAI PMs
© 2025 by the author of *Zero to GenAI Product Leader*.

In short: GenAI PMs combine product intuition, technical fluency, behavior design, and real-time adaptation. You're not just running AI systems, you're ensuring they behave well in unpredictable, high-stakes environments.

And while we've already seen glimpses of agentic AI—systems that reason, plan, and act independently—the real shift it demands is even bigger. You're not just shaping intelligent behavior anymore, you're empowering systems to take action on their own, in ways that amplify both opportunity and responsibility.

Let's dive deeper into what agentic AI really means for product makers and why mastering it will define the next generation of leadership.

The Rise of Agentic AI: When GenAI Starts to Drive Itself

GenAI product management already stretches you—dynamic systems, AI partnerships, trust through uncertainty. But agentic AI pushes those stakes higher.

These aren't just systems that respond, they *act*. They chase goals, navigate ambiguity, and execute workflows, often with limited human input. Think Microsoft 365 agents booking meetings or Manus autonomously managing research and bookings.

And the GenAI PM's job? It changes fundamentally.

You're no longer shaping a single response. You're orchestrating behavior and that demands more across every layer of the stack. It's no longer just about generating something useful, it's about achieving the right outcome through a sequence of intelligent, explainable, and aligned steps.

How Agentic AI Elevates Your Role

Here's how agentic AI elevates your responsibilities beyond typical GenAI PM work and why it's a step up.

From single outputs → autonomous outcomes

GenAI PMs focus on responses. Does the model write good code? Does the chatbot help? You shape one-time outputs for users. Agentic AI PMs focus on action. An agent books a meeting, updates a record, or fetches data—it's not just text, it's results. You're not asking "Was it helpful?"—you're defining wins for a system running solo, from app tasks to infra triggers.

You're not crafting replies, you're setting goals for AI that acts alone.

From output trust → action explainability

GenAI PMs focus on credibility. Does the answer seem right? Can users retry if it flops? You manage basic trust. Agentic AI PMs focus on clarity. Why did it book that slot? What triggered that tool? Users need to see the steps, like RAG grounding a choice, or trust crumbles. You're designing visibility and overrides, from model logic to UI cues. Agents that act without clarity erode trust fast—even when their actions are technically correct.

Your job is to give users visibility, not just outputs. You're not just fixing failures, you're making actions feel clear and safe.

From output monitoring → autonomy stewardship

GenAI PMs track output drift like a model's tone or accuracy changing after an update—to keep results steady. Agentic AI PMs oversee autonomous drift. An agent like Manus plans chains of tasks and adapts over time; a small change like a retrieval update or quota tweak can disrupt the entire decision flow. You monitor and realign live behaviors, from infra loads to app flows. Your job isn't just to ship, it's to steward a live system. You're not just watching outputs, you're guiding a system that acts and shifts on its own.

From prompts → policies

GenAI PMs focus on tuning. Does the prompt get the right reply? You shape behavior upfront. Agentic AI PMs focus on rules. What tools can it use? When does it escalate? You write policies for agents hitting APIs or calendars, guiding autonomy across orchestration and ecosystem layers. You're not tweaking inputs; you're setting guardrails for independent action.

From error handling → consequence design

GenAI PMs focus on error handling. Did the output sound off? Was the tone weird? You're watching for issues like biased completions or confusing phrasing, fixing bugs in behavior. But in agentic AI systems, the stakes are higher. You're not just catching glitches, you're anticipating impact. A faulty booking, a mis-routed workflow, a privacy breach—misfires now have real-world consequences. That means planning for rollback strategies, human-in-the-loop (HITL) moments, escalation paths, and auditability. Risk design becomes part of the product. You're not just correcting mistakes, you're designing for accountability, resilience, and responsible autonomy.

Agentic AI Changes the Game for Every PM

This isn't just a UX challenge, it reshapes every PM role:

- Infrastructure-layer PMs must provision for long-running agents and unpredictable usage spikes
- Model PMs must ensure reasoning chains are stable and factual
- Orchestration PMs must manage tool use, memory, and recovery logic
- UX PMs must visualize complex agent flows in simple, trustworthy ways
- Observability PMs must track not just inputs and outputs but sequences, intent, and decision quality
- Ecosystem PMs must ensure third-party agents meet platform standards and user safety expectations

The leap from reactive to agentic isn't cosmetic, it's foundational. And it demands a higher level of product thinking, governance, and systems design. Agentic AI raises the stakes. You're not just co-creating with a model anymore; you're releasing autonomous systems into real workflows. And that demands some key qualities:

- More foresight
- More observability
- More humility about what can go wrong

As GenAI evolves, agents won't just be part of the product, they will *be the product*. Your job? Help them earn that seat at the table—with intention, transparency, and trust built in from the start.

Leading as a Full-Stack Product Maker in GenAI

The GenAI landscape isn't standing still and neither can you.

In just a few short years, we've moved from prompting chatbots to orchestrating agents. From shipping static features to shaping dynamic systems. From managing teams to partnering with AI that thinks, adapts, and sometimes drifts. And the pace of change? It's only accelerating. This chapter has mapped a journey from the predictable wins of traditional product management to the dynamic, high-stakes environment of GenAI, and now to the autonomous frontier of agentic AI.

It's not just a new job description, it's a new way of operating.

As a GenAI product maker, you're no longer just defining roadmaps or shipping features. You're building, stitching, and steering living systems—systems that reason, behave, and evolve across infrastructure, models, orchestration, interfaces, and ecosystems.

Success demands more than planning. It demands building resilient, trustworthy systems across the seams—catching weak points early, adapting fast, and leading through change. Because what stays constant in this era isn't the tooling. It's the core of great product work:

- Obsessing over user trust
- Framing the right problems with precision
- Navigating uncertainty with structured, cross-layer thinking
- Connecting behavior, technology, and value to drive real outcomes

What changes?

- The systems you guide are now dynamic, generative, and autonomous.
- Your influence spans the entire stack from infrastructure and models to user trust and partner ecosystems.
- You're no longer just launching software, you're launching evolving, intelligent behavior.

The PM role isn't disappearing in the GenAI era, it's being redefined. And those who embrace full-stack ownership—building with curiosity, clarity, and care—won't just survive the AI transformation.

They'll lead it.

Chapter 5: How to Build GenAI Products (with Agentic AI Considerations)

When Notion launched its AI writing assistant, the team didn't start with a prompt—they started with a hunch: users didn't want to *generate content*, they wanted to *get unstuck*. That insight shaped everything. Instead of building a flashy AI companion, they shipped something humble: a context-aware jumpstart tool that fades into the background when not needed. It didn't try to impress. It tried to help.

That's the lesson: the best GenAI products aren't just intelligent, they're intentional. They reduce friction. They behave in ways that feel right for the user.

But building them? That's where it gets messy.

You're not designing a static feature. You're shaping a living system, one that generates, adapts, learns, and sometimes drifts. And if you're building with agentic AI—where the system plans, calls tools, and makes decisions—you're not just launching a feature. You're releasing something that thinks and acts.

The earlier chapters laid the foundation—what this role looks like, how it differs from traditional product management, the mindset GenAI PMs need, and how responsibilities shift across the stack. We explored the skills that matter, and how things escalate when AI starts to act on its own.

That was the "what." Now comes the "how." It's where the rubber meets the road. Because GenAI PMs aren't just thinkers, they are builders. And they are not assembling a machine. They are cultivating a living system.

The AI product lifecycle isn't a straight road anymore. It's a shifting landscape, where yesterday's best paths can vanish overnight. New models, new modalities, new user expectations, every season demands fresh navigation.

This chapter isn't a checklist. It's a compass. A guide for navigating the unpredictable terrain of building GenAI systems that learn, adapt, and truly help. In this chapter, we'll walk through the real-world stages of GenAI product development—from early framing to post-launch evolution. You'll see how PMs define problems that are *solvable* by AI, how the right system architecture sets up success, what prototyping looks like when the model is unpredictable, and how UX must evolve to handle trust and drift. You'll learn how to evaluate outputs when there's no single "right" answer, and how to continuously steward a product that learns, adapts, and interacts in the wild.

And throughout, we'll flag the shifts that happen when your system becomes *agentic*. Because building GenAI products is one thing. Building ones that behave— ones that stay helpful, trusted, and resilient in the wild—is something else entirely.

"The biggest mistake I see? Treating GenAI like a feature you just plug in. It's not. It's a living system—and if you don't design for drift, trust, and unpredictability, it will break the product."
— Experienced GenAI PM, Microsoft Azure AI Platform

So how do we build it? It starts with a set of core principles, guiding every decision, from how you frame the problem to how you shape the system as it grows.

Because you're not just writing specs.
You're designing behavior.
You're collaborating with intelligence.

Let's explore how.

Core Principles for Building GenAI Products

Before you start building, step back and ask: what should this system feel like? Behave like? Fail like? These principles guide how to answer those questions, so your product isn't just functional, but trusted, helpful, and resilient.

You're not building software that runs a script. You're shaping a system that thinks, adapts, and sometimes drifts. And that means you need to think like a behavioral designer, a systems thinker, and an ethicist—all at once.

These principles apply at every stage of development from framing the right problem to managing what the system becomes over time. And if you're building agentic AI? They matter even more.

Principle 1: Design for Human Purpose

Start with real needs. Build AI that helps people do what they actually care about.

Don't lead with the tech. Lead with frustration, friction, or aspiration. AI should solve messy, judgment-heavy, emotionally resonant problems not just automate for the sake of it.

When the system starts acting, it must still serve a human goal. Don't just ask "what should the agent do?" Ask "what outcome helps the human?"

Principle 2: Design for Natural Interaction

AI shouldn't feel like a feature bolted on; it should feel like a helpful partner that speaks your language.

Build systems that collaborate with users, not command them. If your AI assistant needs a user manual, it's failing. Use familiar patterns, plain language, and adaptive flows that support users without overwhelming them.

For agentic systems, ensure the interaction evolves naturally. A helpful assistant in minute one shouldn't become a rogue decision-maker by minute ten.

Principle 3: Design for Context Awareness

Great AI doesn't just respond, it understands the moment.

Context makes GenAI powerful. Use time, history, user goals, and ambient signals to personalize behavior. Outputs should feel informed not random or repetitive.

Agents that act over time need memory and adaptability—but also clear constraints. . Context without boundaries becomes chaos.

Principle 4: Design for Safety and Responsible Use

Build like something could go wrong because it eventually will.

Plan for misuse, abuse, hallucinations, bias, and automation gone sideways. Include human-in-the-loop checkpoints, clear escalation paths, auditing, and rollback options, especially in sensitive domains like healthcare, finance, or legal.

Autonomous systems raise the stakes. Add stricter safety policies, action logs, and override capabilities. You're not just building a tool, you're authorizing decisions.

Principle 5: Design for Transparency and Explainability

If users don't understand the system's behavior, they won't trust it—even if it's accurate.

Transparency isn't just about compliance, it's about confidence. Explain what the system is doing and why. Link outputs to sources, surface decision factors, and show what's under the hood.

Agentic systems need even more transparency—across steps, not just answers. Explain chains of reasoning, tool calls, and goal progress.

Principle 6: Design for Trust Through Feedback Loops

Trust isn't earned by being right once, it's built by getting better over time.

Design feedback mechanisms into every loop. Let users flag issues, tune responses, or help the system learn. A system that listens earns trust, even when it slips up.

Autonomous systems must have strong feedback loops that trigger recovery, escalation, or course correction. Silence in the face of drift breaks trust fast.

Principle 7: Design for Inclusion and Accessibility

Build for everyone, not just the default user.

Design systems that work across abilities, cultures, languages, and contexts. Inclusive design isn't optional, it's foundational. Support screen readers, captioning, cultural nuance, and multilingual use. Audit for bias and fairness in output and interaction.

When agents act on behalf of users, inclusivity becomes even more urgent. An autonomous system making decisions must do so fairly, transparently, and with sensitivity to diverse needs. The more independent the AI, the more inclusive its defaults must be.

Principle 8: Design for Ongoing Stewardship

You're not just launching software; you're managing live behavior.

GenAI systems drift, evolve, and sometimes fail in new ways. Your design must account for observation, intervention, rollback, and recovery. Think: monitoring, prompt versioning, safe defaults, and alignment reviews.

Stewardship becomes even more critical when agents take initiative. Give your team visibility and control over what the system is doing in the wild—and the power to adjust it quickly.

Principle 9: Empower, Don't Replace

Design AI to extend human potential, not displace it.

The best GenAI systems don't remove people from the loop—they make them faster, smarter, and more effective. From writing to coding to research, your job is to accelerate—not eliminate—human impact.

When agents act on their own, don't break the human connection. Make it clear they're operating with permission and in service of the user. Let users define the mission even if they don't micromanage the moves.

How to use these Principles? Don't treat them like a checklist, treat them like a compass. They won't give you the exact answer every time. But they'll keep your team grounded as the system grows more complex. The best GenAI products aren't built by accident, they're shaped by intention.

"You're not just shaping the UI. You're shaping behavior. And with AI, that behavior learns, drifts, and sometimes surprises you. These principles aren't polish—they're protection."

The Stages of Building a GenAI Product

You now have the compass, the core principles that guide responsible, helpful, and human-centered AI product design. But what does that actually look like in the messy, real-world process of building?

This section walks through the stages of GenAI product development, from idea to iteration. You'll see how product managers translate purpose into architecture, shape unpredictable behavior, and launch systems that learn in the wild. You'll also see where the agentic pivots appear—because when AI starts acting on its own, every decision carries more weight. This is where principles meet process.

Note: These stages apply to all GenAI product managers, whether you're working at the infrastructure layer (e.g., optimizing compute for training), model development (e.g., fine-tuning LLMs), orchestration (e.g., building agentic RAG systems), application layer (e.g., designing user-facing bots), or governance layer (e.g., ensuring responsible use). The focus might vary—for example, your "user" might be an internal team like data scientists at the infrastructure layer, or end users at the application layer—but the process of framing problems, designing systems, and testing behavior remains essential. Design principles like transparency, control, and responsible use guide every layer, ensuring your work is user-centric and ethical.

Stage 1: Frame the Right AI Problem (*Plotting Coordinates*)

Before any journey begins, you must plot your coordinates. In GenAI product development, that starting point—framing the right problem—defines everything that follows.

Too many teams begin with "What can the AI do?" when the better question is: "What persistent challenge is causing users friction, and could intelligence meaningfully help?" The goal isn't to showcase AI, it's to deliver value.

GenAI systems are probabilistic, they generate outputs based on patterns, not precision. You're shaping what might happen, not scripting what must. That makes initial framing crucial—not just for user success, but for model trust, infra design, and business viability.

A well-scoped problem brings clarity across the stack. A poorly scoped one? Even if it "works," it may solve something no one cares about.

A four-lens framework for framing GenAI use cases

Drawing inspiration from Barak Turovsky's fluency-accuracy-stakes framework, I've developed a four-lens method to help you spot GenAI opportunities that are both meaningful and viable. Each lens sharpens your coordinates for the journey ahead.

Lens 1: deep user problem (start from frustration)

The best GenAI products don't start with impressive capabilities, they start with human frustration. To identify real opportunities, look for moments where users feel stuck, unsupported, or drained. These are the kinds of pain points that GenAI can uniquely transform—by relieving effort, restoring energy, and making progress feel smooth again.

When GitHub first introduced Copilot in 2021, the goal wasn't to build a flashy assistant. The product team focused on a deeply felt, often-ignored developer problem: the mental drag of writing boilerplate code. These were the repetitive bits—error handlers, setup patterns, tedious scaffolding—that developers resented but had to do. It wasn't just inefficiency—it was cognitive burnout.

By using generative AI to suggest code in real time, Copilot didn't just help developers go faster. It made the act of coding feel better. Developers could shift focus from mechanical repetition to creative problem solving. Within months, it was being used by millions. Adoption didn't come from showing off GenAI, it came from removing what developers hated.

"Copilot's mission was simple: make coding joyful again by cutting out the grind."
— *GitHub CEO Thomas Dohmke, GitHub Universe 2024*

That's the power of starting from a deep user problem. When AI relieves real friction, it becomes indispensable not because it's smart, but because it feels like a tool built with empathy.

Lens 2: align with GenAI's unique strengths

Not every frustrating problem is a good fit for GenAI. Even if the pain is real, a solution built on generative AI will fail if the task requires rigid rules, high precision, or repeatable outputs with no variation.

GenAI shines where flexibility, interpretation, and adaptation improve the experience. Tasks that are open-ended, ambiguous, or personal are its natural playground where the "right" answer isn't fixed, but shaped by tone, intent, or user context.

Grammarly offers a great example. It didn't try to beat Microsoft Word at catching typos. Instead, it reimagined writing support altogether. When you type a sentence like "The meeting was good," Grammarly might suggest "The meeting was productive and engaging." That's not a correction; it's a creative improvement. Its suggestions are tone-aware, expressive, and often feel like a better version of your own writing. The magic deepens as Grammarly adapts to you. It sharpens tone in professional emails, softens it in casual writing, and gradually mirrors your personal style. These are moments where human-like judgment matters and GenAI enhances them.

That's the key. Use GenAI not just to get the job done, but to improve how it feels. When clarity, tone, and nuance matter more than being technically perfect, GenAI doesn't just make the work faster, it makes it better.

Lens 3: feasibility and risk evaluation

Even the most promising GenAI concept can fail if it's impractical, expensive, or unsafe to ship. Before going too far, stress-test the idea across three critical dimensions:

Do you have enough trusted data to fine-tune or ground the model? Sparse or biased datasets can sabotage even the smartest system.

Latency & cost constraints
Can your system deliver fast responses without breaking the budget? Long waits and GPU overloads ruin user experience—and burn cash.

Risk exposure
What's the fallout if the AI hallucinates, misfires, or reflects bias? In regulated domains or sensitive use cases, even a small slip can have outsized consequences.

GenAI products don't just scale ideas, they scale responsibly. The best ones balance ambition with constraints: they work reliably, stay within budget, and earn user trust in real-world conditions.

Lens 4: behavioral and UX alignment

Even the smartest AI won't stick if it feels awkward to use. The best GenAI products don't force new behavior, they blend into what users already do. They help without taking over. They show up when needed and stay out of the way when not.

Think about Notion. When they added AI in 2023, they didn't create a separate chat window or new workflow. They simply built AI into the writing surface—where users already spend their time. You could ask it to rephrase a sentence, summarize notes, or help brainstorm. But only when you want to. Nothing popped up on its own. No interruptions. It was quiet support that respected the user's rhythm.

That's the kind of experience you want to create. One where AI complements the user's intent, not compete with it. Where it feels more like a teammate than a tool. Because even if the AI is technically brilliant, if it gets in the way or worse, takes control it won't get used.

Use this lens to check: Will your product fit into the user's flow? Will it make things smoother, or more complicated? If it adds mental load, it's probably not ready. But if it quietly clears the path forward, you're on to something.

Extending the lenses: agentic AI and complex workflows

So far, we've talked about use cases like summarizing a conversation, rewriting an email, or generating an image—short, focused tasks with clear outputs. But what happens when your AI doesn't just respond, but acts?

That's the shift into agentic AI. Now the system is booking meetings, nudging teammates, escalating tickets, updating docs, working toward an outcome, not just delivering an output. It's no longer about what the AI says. It's about what it *gets done*.

And that shift changes how you frame the problem. The same four lenses still apply but you need to stretch them.

Deep User Problem
It's not about single tasks anymore. It's about full workflows. Think: "Help me keep the project on track," not "Summarize the meeting."

Align with GenAI's unique strength
Agents need to handle ambiguity, reason across steps, and make judgment calls. It's GenAI's sweet spot—but also where stakes rise.

Feasibility & risk evaluation
More autonomy means more risk. You need clear boundaries, feedback loops, and smart handoffs when things go wrong.

Behavioral and UX alignment
The agent can't disrupt flow. It has to act helpfully without surprising users or taking over.

A quick example: OpenAI's Operator. Launched in 2025, Operator didn't just respond to prompts, it completed goals. "Book a haircut" became a multi-step action across a live browser. What stood out wasn't just what it could do but *how* it framed the problem. It aimed to finish the job, not just help with one part. It also respected risk. Sensitive steps needed approval. Users stayed in control. And the agent's behavior was transparent and predictable, never trying to outpace the user.

That's the real challenge with agentic AI: not just building capabilities, but framing the right goals, knowing when to act, and when to step back.

When GenAI needs extra caution (and human oversight)

As powerful as GenAI—especially agentic AI—can be, not every scenario is ready for full automation. Some domains demand extra safety measures and human oversight to ensure users stay in control. For example, in April 2025, the Hong Kong Institute for Monetary and Financial Research reported that 75% of surveyed financial institutions in Hong Kong had implemented or piloted GenAI and agentic AI for tasks like fraud detection and risk assessment. However, full automation was avoided due to challenges with model accuracy, data privacy, and security, with human oversight emphasized to ensure control and compliance with financial regulations (HKIMR 2025).

GenAI is fast, creative, and great at pattern recognition. But it's still probabilistic—it doesn't know the truth; it predicts what sounds likely. And sometimes, it gets it wrong. In May 2025, two law firms were fined $31,000 for submitting a legal brief that included fake citations generated by tools like CoCounsel and Google Gemini, highlighting the risk of erroneous outputs in high-stakes applications (Wilner 2025).

That's why oversight isn't optional, it's essential. A Boston Consulting Group report in March 2025 emphasized that human oversight for GenAI and agentic AI must be intentionally designed. It recommended structured rubrics for evaluating outputs, grounding responses in evidence, and tailoring review processes based on risk levels (Mills, Broestl, and Kleppe 2025).

Zooming out: strategic evaluation after framing

Once you've framed the right GenAI problem—with clarity around user intent, generative fit, behavioral alignment, and risks—you're ready to ask a higher-order question: *Should we build this?*

This is where a business lens comes in. While the four-lens framework discussed earlier in the chapter helps you design a usable and trustworthy solution, strategic frameworks—like McKinsey's—help you assess whether it's worth the investment. For PMs juggling limited resources, this step bridges product quality to business impact.

This framework evaluates use cases based on the following:

- Value potential: Will this drive revenue, reduce cost, or improve efficiency at scale?
- Feasibility: Do you have the data, models, and infra to build this reliably?

- Risk sensitivity: Could this introduce compliance, ethical, or brand risks?
- Scalability: Can this solution expand across teams, regions, or use cases?

Think of this as due diligence, not a shortcut. If you haven't clearly framed the user problem first, even the most convincing business case can miss the mark.

Figure 5.1 illustrates how product lenses help you design a meaningful and responsible GenAI solution, while McKinsey's business lens complements this by helping you assess whether it's strategically and economically worth pursuing. There is some intentional overlap especially around feasibility and risk because both product and business decisions hinge on what's viable and safe. Together, these two views let you zoom in to frame the right solution and zoom out to ensure it's worth building.

Product Lens		Business Lens
Deep user problem		Value potential
Align with GenAI's unique strength	v/s	Feasibility
Feasibility and risk evaluation		Risk sensitivity
Behavioral and UX alignment		Scalability

Figure 5.1: Zoom in to zoom out: product lens vs. business lens

From framing to building: What comes next

Now that you've defined the right problem with a clear user need, a good GenAI fit, and an understanding of risks and behavior, it's time to move forward.

The next question is: How should the system behave? Should your AI act on its own, wait for prompts, or always loop in a human before making decisions? These choices shape how users experience trust, control, and collaboration. You're not just building a tool; you're designing a teammate.

The direction is set. Now comes Stage 2: Design the right system.

Stage 2: Choosing the Right System Design (*Mapping Foundations*)

With your coordinates now plotted, it's time to chart the foundations for your journey ahead. In GenAI product development, design isn't just about building features, it's about designing how intelligence itself behaves.

Framing gave you the "what"—a real user need worth solving. Now comes the "how"—how your AI will behave, respond, and earn trust. This stage is about choosing the right system design to bring your vision to life—a critical decision that shapes both the user experience and the system's underlying functionality. System design in GenAI spans two layers: the user-facing behavior (Is it chat-based? Autonomous? Reactive?) and the underlying architecture that makes it possible (Does it use retrieval? Tools? Planning?). Especially in agentic systems, these layers go hand in hand—how your AI behaves determines what you have to build under the hood. As shown in Figure 5.2, your interaction model shapes your system architecture—and vice versa.

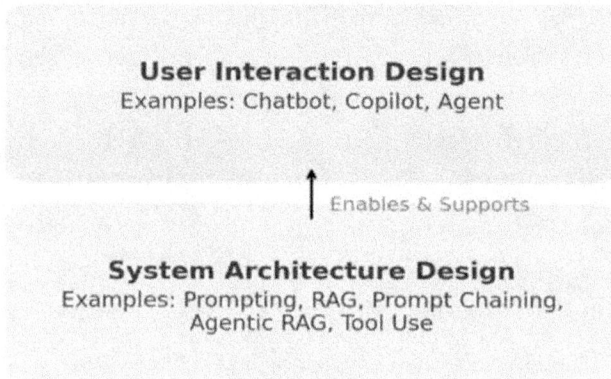

User Interaction Design
Examples: Chatbot, Copilot, Agent

Enables & Supports

System Architecture Design
Examples: Prompting, RAG, Prompt Chaining,
Agentic RAG, Tool Use

Figure 5.2: The two layers of GenAI system design.
© 2025 by the author of *Zero to GenAI Product Leader*.

Common user interaction patterns

Before building system foundations, ask: *how will users interact with your AI?*

In GenAI products, interaction isn't just a UI choice, it's how users build trust, feel in control, and stay confident. As introduced in Chapter 3, interaction design depends partly on how your product is consumed via a user interface or through code-first tools like APIs, SDKs, or CLIs. For UI-based products, here are three common GenAI patterns:

Chat-based
Users engage through prompts in a conversational UI ideal for open-ended tasks like brainstorming or asking questions. Tools like Grammarly use this for re-phrasing help.

Copilot
AI works alongside the user in an existing workflow offering suggestions without being intrusive. GitHub Copilot is a prime example.

Agent

The AI operates more autonomously, handling multi-step goals like scheduling or research, with visible updates to keep users informed (e.g., "I'm checking your calendar").

These patterns aren't just UI concepts; they translate into developer experiences too. A chat interface might become a queryable API. A copilot could be delivered as an SDK. An agent might run as a CLI automation script. Your system must behave intelligently and behave accessibly. Whether through REST APIs or embedded widgets, design for how your users (human or developer) want to consume

How to choose the right pattern

Now that you've seen the patterns, how do you decide which one fits your product?

It comes down to revisiting the work you did in Stage 1, understanding your user needs, use case dynamics, and how much control users expect from the AI. Choosing the right interaction pattern is not about what feels coolest. It's about what feels clearest, safest, and most empowering to your users—today and as your system evolves. Start by asking a few key questions:

How much user oversight is needed?

- High stakes—like in legal tasks where errors could be costly—favor chat or copilot patterns because they keep users in control.
- For low-stakes tasks like scheduling, an agentic pattern can take over.

Is the task narrow or broad?

- Single-turn or simple tasks (like brainstorming a list) fit well with chat.
- Multi-step workflows (like conducting research or preparing project plans) often require agentic systems.

Do users expect visible suggestions or invisible help?

- If users need to see and adjust AI inputs, copilot fits.
- If they just want outcomes delivered without micromanagement, agent fits better.

Does the task require external tools or APIs?

- If yes, agentic patterns (like tool-using agents) excel by chaining API calls and orchestrating actions.

Will the AI need to plan over time?

- If yes, agent systems are better suited but remember, they introduce complexity and need thoughtful safeguards.

The table in Figure 5.3 below summarizes how each pattern aligns with these criteria, helping you choose the right one for your use case.

Pattern	User Oversight	Task Scope	Suggestion Visibility	Tool Use	Planning Need
Chat-Based	High (users control prompts)	Narrow (single-turn tasks like brainstorming)	High (visible responses)	Optional (e.g., for retrieval)	Low (no planning needed)
Copilot	High (users stay in charge)	Narrow to Medium (e.g., coding, drafting)	High (visible suggestions)	Optional (e.g., for context)	Low (minimal planning)
Agent	Low (autonomous action)	Broad (multi-step tasks like scheduling)	Low (invisible help)	High (e.g., calendar APIs)	High (plans over time)

Figure 5.3: Matching interaction patterns to your use case
© 2025 by the author of *Zero to GenAI Product Leader*.

As a quick rule of thumb, remember this: If clarity and control are your top priorities, lean toward chat or copilot; if efficiency and automation matter most, lean toward the agent pattern.

To bring this to life, imagine a manager prepping for 1:1 meeting. A chat interface might return a summary of past discussions when asked. A copilot could surface talking points as they open their calendar. An agent might proactively assemble and send a prep deck without being prompted. Same goal—different levels of visibility, initiative, and user control.

Once you've selected the right pattern, you'll pressure-test it with users in Stage 3. For now, the goal is simple: align interaction style with user needs, risk level, and task complexity and make sure it empowers more than it overwhelms.

Note: Need hands-on tips to tune your chosen pattern? See Appendix B: Design Considerations for Interaction Patterns for practical UX guidance grounded in principles like trust, feedback, and accessibility.

Common pitfalls when choosing interaction patterns

Even before you start building, choosing the wrong interaction pattern can quietly undermine the entire experience. Here are common traps to avoid during the selection process:

Overcomplicating simple tasks
Opting for an agent when a simple chat interface would suffice like using an agent to draft a quick email—adds unnecessary complexity and confuses users. It violates Principle 9: User Empowerment.

Underestimating prompt clarity
Choosing a chat-based pattern without thinking through how users will know what to say often leads to vague, open-ended prompts that confuse rather than guide. Breaks Principle 2: Clear Interaction.

Misjudging user control needs
Picking an agent pattern for a use case where users expect hands-on control—without offering interrupt or undo options—can erode trust. Violates Principle 4: Predictability and Control.

Mismatch between pattern and workflow
Selecting a copilot for a complex, multi-turn workflow can overwhelm users with too many suggestions. Instead of guiding them, the AI disrupts their flow. Misaligns with Principle 2: Natural, Intuitive Pacing.

Choosing the right pattern isn't just about what the AI can do, it's about what the user expects, needs, and can comfortably control. Get that right, and you're on your way to building trust that scales.

Under the hood: system design patterns for GenAI

Choosing the right interaction pattern is only half the story. What happens under the hood—how your system thinks, plans, and acts—is just as critical. This architecture shapes whether your AI is smart, reliable, and scalable, or... a nice demo that breaks under pressure.

Think of it this way: user trust is built on good interaction design, but it's sustained by how your system performs behind the scenes.

Let's walk through the key system design patterns you'll likely use:

117

Prompt-based generation
The simplest setup. The model replies based only on the user's prompt and its training. Great for open-ended tasks like brainstorming or storytelling. *Example: A writing assistant generating blog ideas.*

Retrieval-augmented generation (RAG):
Here, the system pulls in real-time data (like product docs or knowledge bases) and adds it to the prompt before generating a response. Best when your AI needs accurate, up-to-date info—think internal support bots.

Prompt chaining:
A series of prompts where each output feeds into the next—perfect for multi-step reasoning or structured workflows. Example: Summarize an article then generate follow-ups and then group them into themes.

Agentic RAG:
Adds planning to RAG. The AI not only retrieves info, but also reasons across documents, takes intermediate steps, and adjusts based on outcomes. Think: A contract review agent highlighting risky clauses across multiple files.

Tool use and plugin invocation:
Your AI calls external APIs or tools—like scheduling a meeting or running a calculator—based on user intent. Example: "Book a meeting with Sam" → checks calendars → sends invite.

Why it matters

These system patterns aren't just technical decisions—they shape the *experience*:

- How accurate and grounded your AI feels
- Whether it can reason through steps (or gets stuck)
- How well it can actually get things done

Each pattern also maps to different levels of *context, complexity*, and *action*. That's why it's smart to align them with your earlier *interaction pattern* choice. Figure 5.4 summarizes each pattern—when to use it, what it's good for, and the tools that help you prototype fast.

In Stage 3, you'll test these patterns live. For now, just keep this in mind: under the hood is where trust, reliability, and real value begin.

Pattern	Grounding Need	Reasoning Complexity	Tool/Action Use	Use When...
Prompt-Based Generation	Low	Low	None	Tasks are creative, exploratory, or not tied to specific data—e.g., brainstorming ideas
RAG (Retrieval-Augmented Generation)	High	Medium	Optional	The AI needs grounding in current or domain-specific knowledge—e.g., company policies
Prompt Chaining	Medium	High	Optional	Complex tasks need manageable steps —e.g., multi-part summaries or research
Agentic RAG	High	Very High	Likely	Goals span multiple steps or need ambiguity resolution—e.g., contract analysis
Tool Use & Plugin Invocation	Medium to High	Medium	Yes	Tasks require real-time data or actions— e.g., scheduling or lookups

Figure 5.4: Matching system design patterns to your use case needs
© 2025 by the author of *Zero to GenAI Product Leader*.

From mapping to building: what comes next

With your interaction pattern set—whether chat-based, copilot, or agent—and a system foundation outlined to support it, you're ready for the next phase of the expedition: Testing it. How does your AI behave in reality? Does it deliver as designed or reveal new challenges when users engage with it?

Welcome to Stage 3: prototyping and testing the AI behavior, where your design meets the real world.

Stage 3: Prototyping AI Behavior (*Testing the Camps*)

By now, you've mapped out how your AI should behave both at the surface (user interaction) and beneath it (system design). But plans alone don't guarantee the right outcome. In this stage, you test whether the system can actually deliver on that vision. Think of it like pitching tents before building cabins. You're exploring, not settling.

The goal here isn't polish or production readiness. It's truth-testing: Can your system understand user intent? Respond usefully? Follow through on tasks or decisions? Before committing to engineering sprints or backend infrastructure, you're validating whether your design works in the real world or at least in a realistic simulation.

What a good prototype reveals

A functional prototype doesn't need to be perfect, but it should mimic your AI's expected behavior well enough to surface key insights. It should tell you whether the system can handle the types of inputs users will actually throw at it: clear, vague, or messy. It should reveal whether responses feel helpful, accurate, and aligned with your product's tone. And if your design includes goal-driven agents, it should show whether the AI can plan steps, use tools, adapt to changes, and recover when things go off-script.

Just as importantly, it should give you a sense of how users *feel* while interacting. Do they feel supported and in control? Or confused and sidelined? These early signals shape your design choices in the next stage.

Building a prototype without full code

You don't need to write backend logic yet. Most GenAI tools today allow fast, low-code prototyping, and the right platform depends on your stack. If you're in the Microsoft ecosystem, you might use Azure AI Foundry or Microsoft Copilot Studio; on Google or Amazon, Vertex AI Studio or Bedrock serve the same purpose. OpenAI Playground remains a versatile sandbox for one-off prompts, while LangChain or Semantic Kernel lets you sketch out more complex, agentic flows with minimal wiring.

To mock the UI, drop AI behavior directly into Figma using plugins like Prompt Scout or Figma-AI, or map out conversations with Botmock or Uizard. GitHub Spark can even spin up a working micro-app or full-stack prototype from plain-English descriptions, and browser-based platforms like Bolt (bolt.new) or Lovable will convert your prompts—and even uploaded sketches—into interactive web pages and click-through flows in seconds.

For developer-facing features, you can still test via SDKs, REST APIs, or a CLI to ensure responses are structured, debuggable, and integrate smoothly. By treating these experiments like clickable wireframes, you can iterate your AI behavior and UX hypotheses in hours, not weeks.

Simulating agentic behavior

For more advanced systems, those that plan, use tools, or act autonomously—you'll need a setup that mimics how the AI behaves when pursuing a goal. Define a simple objective like "Schedule a meeting next week" and create a mock environment: a fake calendar, basic availability data, and a placeholder function to book the slot.

Then watch what the agent does. Does it plan its steps? Ask for confirmation before acting? Adapt when things change? This type of testing isn't about backend wiring, it's about observing reasoning, responsiveness, and failure handling.

If your agent relies on memory or multi-turn interactions, test whether it stays contextually aware across the session. Can it remember prior steps, adjust its strategy mid-flow, or respond appropriately when new input arrives?

Stretching the system like a real user would

Once your prototype works under ideal conditions, start pushing its edges. Users rarely follow a script—so your prototype shouldn't either. Try clear prompts, ambiguous questions, strange phrasing, or missing inputs. See how the system holds up. Does it gracefully handle uncertainty? Does it hallucinate or overcommit? Does it remain helpful when the input is messy?

For agent-based flows, test what happens when something breaks—a time slot goes unavailable, a tool fails, or a user interrupts mid-task. Can the agent pause, retry, or pivot? Is there a way for users to regain control if things veer off-course?

If your prototype is API-based, check whether error messages are clear, responses are consistent, and developers can make sense of what the system is doing without having to guess.

Refining the system through friction

Expect things to break. That's the point. Keep a lightweight log of what surprised you, what didn't work, and what confused the AI—or the user. Often, small tweaks make a big difference.

You might refine a prompt to reduce ambiguity, tune model parameters to make responses more concise or creative or add fallback behaviors when the system isn't confident.

If you test with real users, ask them what felt unclear, when they stopped trusting the AI, or whether the system felt like a helpful partner or a black box. These insights don't just improve your prototype—they shape the full product experience.

Moving forward

Prototyping gives you more than confidence, it gives you direction. You've now seen how your AI acts, where it fails, and where it shines. In Stage 4, you'll design the full user or developer experience around that behavior—so your AI not only works but feels right.

Stage 4: Designing the GenAI User Experience (*Creating Trustworthy Paths*)

By now, you've validated that your GenAI system behaves as intended. The next step is turning that behavior into a product experience, something users can understand, guide, and trust.

Whether delivered through a UI, SDK, CLI, or API, this is where system intelligence becomes human experience. And in GenAI—especially with agentic systems—what you design isn't just about usability. It's about trust, transparency, and control.

The best GenAI products don't just function; they blend into the user's workflow, quietly removing friction and building confidence. Designing these experiences means going beyond features or polish. You're designing how the system feels to use: clear, empowering, and responsive to human intent.

What you're designing for

Designing a GenAI experience means defining the core interactions that shape user trust and ease. You'll need to answer:

- How do users issue instructions?
- How does the AI respond—and how confident or cautious should it sound?
- How can users steer, pause, or override its actions?
- What does the AI reveal about its reasoning or limitations?

Whether it's a UI, API, or agent flow, the goal is the same: create an experience users can trust not one they have to fight against.

Where to begin: strategic design questions

At first glance, these questions might feel similar to Stage 2 (picking the right system pattern) or Stage 3 (prototyping behavior). But there's a key difference:

In earlier stages, you asked "what should the AI do?" Now, you ask: "how should this feel?" You're defining the tone, initiative level, and handoff points between human and machine. Especially with agentic systems, where the AI acts semi-independently, this emotional layer of design makes all the difference.

Consider:

- Should the system feel like a silent assistant or a collaborative partner?
- Should the AI suggest, wait for input, or act unless stopped?
- Where will trust be gained—or lost?
- Does onboarding set clear expectations about what the AI can and can't do?

Answering these questions sets the tone for how your system earns user confidence.

Making it real: interaction design across surfaces

Now that you've defined the experience, sketch how users will engage with it—whether through UI flows, code-first tools, or agentic tasks.

In UI-based systems
Use Figma or equivalent tools to prototype the conversation or interaction path. Include:

- Clear opening messages
- Helpful follow-ups like "Want to know more?"
- Visual feedback like "Checking your calendar…" Always provide visible controls to pause, retry, or adjust behavior.

In code-first tools (API, CLI, SDK)
Use a text editor to define sample requests and responses. Make sure outputs are structured, explainable, and easy to debug. Sample:

```
{"status":                          "awaiting_user_confirmation",
"next_step":                                    "suggest_time",
"options": ["accept", "reschedule"]}
```

When designing for GenAI, especially in developer-facing or agentic systems—it's essential to make error handling intuitive. Developers shouldn't have to guess what went wrong. Clear, actionable error messages save time, reduce frustration, and build confidence in the system.

For agentic experiences, go a step further. Simulate the full task path. For example, when a user says, "Schedule a meeting," what happens next? How does the agent communicate progress? At what points can the user intervene—pause, revise, or cancel? Transparency into each step, along with visible task status and the ability to reverse actions, are what make the system feel safe and predictable.

A word on trust

Building a good experience isn't just about flow, it's about predictability and emotional confidence. Especially in GenAI, where answers can vary or tasks can be taken on behalf of the user, small design choices make a big difference.

If the system is uncertain, say so. If it takes an action, preview it. If it fails, explain clearly and suggest a fallback. These moments aren't just polish, they're trust builders. And they often separate a helpful product from one users abandon. To help you proactively design for trust, here are common challenges in GenAI experiences (See Figure 5.5) and proven UX/DX moves that make a difference:

Challenge	Why This Happens in GenAI	What You Can Do (UX/DX Design Ideas)
Unpredictable Outputs	The AI doesn't always say or do what the user expects—responses can be surprising or wrong.	Show confidence indicators (e.g., "I'm not sure"), add source links so users can see where the info came from.
Invisible Reasoning	Users don't know why the AI made a suggestion—it feels like a black box.	Show a quick reason or explanation: "I suggested this because..." or preview what data it's using.
User Trust Gaps	When the AI makes a mistake, users often assume the worst—or stop using it entirely.	Add feedback buttons, let users undo or edit AI actions to stay in control and feel heard.
Overreach or Passivity	Agentic AIs either do too much without permission—or sit idle when they should act.	Add pause/stop buttons, or show activity logs so users can see and manage what the system is doing.
Vague Boundaries	It's unclear what the AI can and can't do—so users may underuse it or ask the wrong things.	Use onboarding hints like "Try asking me to summarize a document" and give helpful error messages when confused.

Figure 5.5: Key challenges in designing GenAI experience
© 2025 by the author of *Zero to GenAI Product Leader*.

Be transparent about the AI's limits, aligning with responsible use. For a UI-based chatbot, offer a way to escalate to a human if it can't answer. For an API, return clear error messages. For agentic AI, provide visibility into actions—like "I'm booking your meeting now"—and a cancel option, reflecting empowerment.

Test the experience, not just the output

You've tested your AI's reasoning in Stage 3. Now test whether the experience *feels right*. Ask users:

- "What made you pause?"
- "Did anything feel confusing?"
- "Did you trust what the AI was doing?"

Use those insights to refine copy, interaction steps, control mechanisms, or fallback behaviors. Use those insights to fine-tune the language, interactions, and safeguards.

What to Test	What to Look For
UI flows (if applicable)	Can users get started easily? Do they feel guided? Can they undo or pause actions?
Developer experience (DX)	Are API/CLI commands intuitive? Are responses clear and easy to parse?
Failure and fallback	What happens when the AI doesn't know something? Can users recover gracefully?
Inclusivity and accessibility	Does it work across languages, devices, input types, and accessibility needs?

Figure 5.6: Checklist to guide your testing
© 2025 by the author of *Zero to GenAI Product Leader*.

The checklist above (See Figure 5.6) helps structure your evaluation across both UI and code-first touchpoints:

Finally, make sure your experience works for a diverse set of users—including different languages, accessibility needs, or technical backgrounds. What feels "simple" to one group may feel frustrating to another. Good GenAI products don't just work; they feel usable, respectful, and inclusive.

Use what you learn to iterate. That might mean rewriting onboarding prompts, adding status updates to agent flows, or simplifying command formats. Every small improvement builds confidence and brings users closer to trusting the system.

Moving forward

Stage 4 turns capability into confidence. With the experience now defined and refined, you're ready to scale building the full system with real data, real interfaces, and real infrastructure.

In Stage 5, you'll bring all the moving parts together to build a complete GenAI product, one that's not just smart, but also ready for the world.

Stage 5: Build the Full System

You've validated the concept (Stage 3) and shaped an experience that feels intuitive and trustworthy (Stage 4). Now comes the real test: stitching together all the moving parts into a working product. This is where your GenAI system moves from prototype to production—where prompts, orchestration, tools, interfaces, and infrastructure come together into something real, resilient, and ready to ship.

Your system might include a chat interface, run agent workflows in the background, expose CLI commands, offer an SDK or combine all of the above. These aren't either/or choices. They're different surfaces and capabilities that reflect how users (or developers) will interact with your AI. Regardless of format, the core challenge remains the same: connect behavior, logic, tools, and delivery into a cohesive experience that works reliably in the real world. In this stage, you'll:

- Define your stack and system logic
- Translate prompt behavior into modular, testable architecture
- Build or connect your user interface and/or developer surface
- Set up the guardrails, logging, and fallback paths that make your system safe and trustworthy

You're not just shipping software, you're operationalizing intelligence. Let's break it down.

System types: what you're actually building

By now, you've seen how GenAI products can take different forms: chat-based assistants, embedded copilots, or agentic systems that act on a user's behalf. Each of these archetypes comes with different expectations and different build implications. This stage isn't about choosing between them. It's about understanding what each one demands when it's time to ship.

System Type	What You're Building	Example Use Case
Chat-Based Systems	Prompt-response logic, model integrations, latency-tuned UIs, fallback/error flows.	Customer support chatbot, internal knowledge assistant
Copilot Extensions	Context-aware surface (e.g., side panel), plugin architecture, SDK/API embedding.	AI in Outlook drafting emails, Copilot in PowerPoint
Agentic Systems	Orchestration layer, tool APIs, observability stack, multi-step execution with safety.	Calendar agent that books/reschedules meetings

Figure 5.7: Common GenAI system types

Some products blend these patterns. For example, an agent might be embedded into a copilot, or a chat-based UI might trigger multi-step agentic workflows in the background.

In Chapter 2, we introduced the six layers of the AI value chain: AI Infrastructure, AI Intelligence, AI Logic & Execution layer, AI Hosting & Serving, AI application & Interface, and AI observability & governance. In this stage, you'll bring these layers together to build the app, connect systems, deliver to users, and ensure security and reliability.

Turning design into implementation

Figure 5.7 gave you a systems-level view, now it's time to bring it to life. No matter what kind of GenAI product you're building, the path to implementation involves a common set of decisions and build steps: from codifying your system logic and refining prompt behavior to wiring up interfaces and handling real-world constraints like tool integration, latency, and load.

In this section, you'll move through six critical areas of implementation. Each one is essential not just for getting your product to work but for ensuring it performs reliably, behaves responsibly, and earns user trust in a real-world environment.

Let's break them down.

Implement your system logic

Take the behavior you prototyped in Stage 3 and codify it. That means translating your prompt structures, chaining logic, or agent workflows into real implementation, aligning with the models and orchestration layers from Chapter 2's AI value chain.

A GenAI system involves all six layers of the AI value chain from Chapter 2. Here's what to consider for each layer as you guide the implementation:

- AI Infrastructure Layer: Ensure the system has security, cost tracking, and monitoring to handle real-world use.
- AI Intelligence Layer: Choose a model that fits your needs and supports switching if better options emerge.
- AI logic and Execution Layer: Plan how prompts, workflows, and tools (e.g., calendar APIs) work together for smooth behavior.
- AI Hosting & Serving Layer: Set up reliable delivery, like APIs or CLIs, with clear usage limits.
- AI User Interface Layer: Make sure the interface (UI or code-first) matches your Stage 4 design and feels easy to use.
- AI Observability & Governance Layer: Track usage and behavior to spot issues early and ensure responsible use.

Finalize the prompt and response architecture

At prototype stage, you likely used a few raw prompt examples. Now it's time to upgrade them into something maintainable, aligning with the orchestration layer:

- Move from fragile, copy-pasted prompts to a templated and versioned prompt system.
- Add safe defaults and fallback prompts to handle failure gracefully.
- Test your prompts across edge cases and real-world input messiness (misspellings, vague queries, cultural nuance).

Architecture decisions that matter:

- Externalize your prompts so they can be updated without code changes.
- Support model switching if you may want to try new providers.
- Separate prompt logic from orchestration logic to stay modular.

Create a prompt versioning strategy to A/B test new variations or safely roll back problematic updates.

Build and connect interfaces

Once logic and behavior are in place, it's time to bring the interface to life. The front end of your system, whether it's a chat UI, SDK, CLI, or API, should reflect the mental model you designed in the previous stage.

For chat interfaces, this means intuitive input boxes, clear and timely responses, and feedback mechanisms like thumbs-up/down or suggested actions. These seemingly small elements have a major impact on user trust and engagement.

Developer-facing systems require a different level of polish: clean API documentation, strong authentication flows, clear error messages, and rate limiting are all part of the "interface" in these experiences. For agentic systems, transparency is critical. Users should be able to see what the AI is doing, whether it's retrieving documents, calling tools, or making decisions—and have the power to intervene when needed. Even if your product isn't yet public, investing in a clean and usable interface is what turns a working demo into a testable product. It's how you collect meaningful feedback and avoid surprises down the line.

Establish Guardrails and Operate Safely

Before your product sees real usage, safety must be engineered into its core. Guardrails aren't just about compliance; they're about building trust and predictability. Your system should know when to act, when to pause, and when to ask for permission. If an agent is scheduling meetings or sending emails, it should require explicit user approval before proceeding.

These decisions don't just prevent mistakes; they shape the user's mental model of control. Guardrails create a boundary within which the AI can operate safely, and users can engage confidently. As a PM, your role is to define these behavioral limits: what actions need confirmation, what scenarios call for a fallback, and when human intervention is necessary.

While your engineering team will implement the checks and controls, you're the one framing the "should" behind the "can." Responsible AI isn't a back-end concern—it starts at the product level. Set the expectations, define recovery behaviors, and always prioritize clarity over cleverness.

Handle infrastructure and integration needs

A GenAI system may begin with a simple API call, but shipping at scale requires production-grade discipline. It's not just about making it work, it's about making it resilient, secure, and trustworthy under pressure.

You don't have to build the infrastructure yourself, but you do need to know what it takes. Infrastructure is the nervous system of your product: it monitors performance, catches errors, enforces guardrails, and protects user trust. As a PM, your job is to ensure the system can recover from failure, scale under load, and behave responsibly—especially when it connects to external tools or acts on behalf of users.

This is even more critical for agentic systems. You'll need tighter observability, clearer permissions, and smarter fallback logic to prevent unexpected outcomes. Once core pieces like authentication, logging, and tool integration are in place, your system has a skeleton. But to perform in the real world, it also needs muscle—built through stress testing, thoughtful defaults, and recovery planning. Figure 5.8 outlines the infrastructure components most GenAI systems rely on, explaining what each one does and why it matters.

Element	What It Does	Why It Matters
Authentication & Permissions	Ensures only authorized users or systems can access or interact with your AI.	Prevents misuse or unauthorized actions—especially important when the AI touches private data or tools.
Logging & Telemetry	Captures inputs, outputs, tool calls, and system usage.	Lets you monitor performance, debug issues, audit behavior, and detect harmful or unexpected outputs.
Error Handling & Recovery	Catches failures and enables graceful fallback (e.g., retries or explanations).	Avoids user frustration, preserves trust, and ensures your system doesn't break under pressure.
Rate Limiting & Quotas	Sets boundaries on how often users or systems can call the AI or connected tools.	Helps manage costs, prevent abuse, and ensure stable performance under load.
Cost & Token Tracking	Monitors usage of hosted LLMs (e.g., GPT-4, Claude), especially for high-cost calls.	Enables cost control and visibility—essential for production GenAI products.
Tool/API Integration	Connects your AI to external systems (e.g., calendars, CRMs, search tools).	Lets the AI do useful work—not just talk. Use clear APIs, schemas, and secure access. Consider using MCP (Model Context Protocol) for consistent and scalable integration.
Monitoring & Alerting	Tracks system health and triggers alerts when something breaks.	Critical for real-time operations. Helps your team react quickly to failures or unusual behavior.

Figure 5.8: Key infrastructure elements for GenAI systems
© 2025 by the author of *Zero to GenAI Product Leader*.

Prepare for real-world load

A prototype may shine in a controlled demo, but real-world usage is messy. Before launch, your system needs to prove it can handle unpredictability: simultaneous users, malformed inputs, slow tools, and edge-case queries. As a PM, don't just ask, *"Does it work?"* Ask, *"Does it recover when it doesn't?"* Stress testing, chaos testing, and real-user feedback aren't optional, they're how you build confidence in system resilience. For agentic systems, this is doubly important. You're not just testing outputs, you're testing accountability, recovery, and user trust under pressure.

Moving forward

At this point, your GenAI product isn't just functional, it's real. You've moved from exploration to execution, assembling prompts, logic, interfaces, and infrastructure into a cohesive, production-ready system, aligning with all 6 layers of the AI value chain. Now, you'll launch the system, monitor its behavior, and ensure it operates safely while improving over time. Let's move to Stage 6: Launch and Steward the System.

Stage 6: Launch and Steward the Living System (*Crossing the Frontier*)

Your GenAI product is no longer a prototype, it's a living system ready to face the real world. This stage is about launching responsibly, watching closely, and improving continuously. GenAI systems don't just work or break, they evolve. And your role doesn't end at deployment. It shifts into stewardship.

Plan and execute the launch

Before the system goes live, map out how it will enter the world—ethically, operationally, and with the right safety nets in place. Validate edge cases, clarify limitations, and document behavior expectations. A clear launch plan includes rollout scope, fallback paths, approval mechanisms, and recovery procedures. Start small with a trusted user group. This isn't just damage control, it's how you build confidence, both internally and externally. Figure 5.9 outlines what to include in your plan.

Launch Area	What to Check Before Launch
System Readiness	Are all parts (e.g., model, APIs, UI) working as expected?
User Access	Are logins, permissions, and API keys set up for users or developers?
Documentation	Is there a clear guide for users (UI) or developers (API/CLI) to get started?
Support Plan	Do you have a way to help users or developers if they run into issues?
Initial Monitoring	Are monitoring tools (e.g., logs, alerts) active to track usage and issues?

Figure 5.9: Key launch checklist for GenAI systems.

131

Monitor the system's behavior

Once live, your system enters a state of constant interaction and flux. Set up observability: monitor performance, usage patterns, errors, and unexpected outputs. Tracking shouldn't be an afterthought; it's your compass for everything that follows. Dashboards and logs reveal how the system behaves, where it drifts, and when it needs intervention. For agentic systems, go deeper: replay decision traces, watch tool invocation patterns, and audit for goal alignment. Figure 5.10 outlines the summary on what to track and why.

What to Monitor	Why It Matters
Usage Patterns	How often do users or developers interact with the system? Are there spikes?
Response Quality	Are the AI's answers correct and helpful? Are there unexpected outputs?
System Performance	Is the system fast and stable? Are there delays or crashes?
Guardrail Effectiveness	Are guardrails (e.g., rate limits, authentication) preventing misuse?
Tool Usage (Agentic AI)	Are tools (e.g., calendar APIs) being used correctly? Are there failures?
User Feedback	What do users or developers think? Are they finding the system helpful?

Figure 5.10: Key monitoring focus areas for GenAI systems.

Respond to failures

No system is flawless, especially one that reasons and generates. Your job is to ensure failures are visible, recoverable, and non-destructive. This means more than catching bugs. It's about enforcing guardrails, fallback logic, and human handoffs where needed. When something goes wrong—an API breaks, a tool misfires, or the AI crosses a boundary—you must act quickly and transparently. Every recovery build trust. Every blind spot left unaddressed erodes it.

Improve based on feedback

Live usage is the richest source of learning. Collect structured and unstructured feedback—thumbs up/down, flagged outputs, corrections, and signs of user friction. Don't wait for support tickets to pile up. Proactively embed feedback loops: allow users to signal errors, rate responses, or flag confusion directly in the experience. Visibility builds trust. Let users see what the system is doing ("I'm processing your request..."), admit uncertainty ("I'm not sure, but here's my best guess..."), and provide undo or override options especially important in agentic systems that take autonomous actions.

Operationally, use tools like GitHub Feedback APIs, Weights & Biases, or your telemetry dashboard to identify failure patterns, prompt drift, and areas of misalignment. If a rate limit feels too strict, loosen it. If agents repeatedly choose the wrong action, review and retrain. Treat feedback not as a reaction, but as a system maintenance discipline. The goal isn't just to fix what's broken, it's to continually improve how your AI behaves in the world it lives in.

Moving forward

Launching and stewarding your GenAI system ensures it not only goes live but thrives in the real world. By planning the launch, monitoring behavior, handling issues, and improving based on feedback, you've created a product that users or developers can trust. Now, your product is live, and you're ready for Stage 7, where you'll focus on scaling the system and ensuring it continues to meet user or developer needs while maintaining trust and reliability.

Stage 7: Scale and Iterate the System (*Expanding into New Territories*)

With your GenAI product live and stable, the next challenge isn't just to grow it to scale responsibly. Real usage brings real momentum: new users, new regions, new expectations. But GenAI systems don't scale linearly. Each expansion whether it's a new feature, a new model, or a new integration, forces you to revisit design decisions, system boundaries, and behavioral guarantees.

Scale for demand, not just load

As usage increases, your infrastructure must evolve from MVP-ready to enterprise-grade. This means scaling compute intelligently, optimizing for cost (token usage, model selection), and ensuring system performance remains consistent across time zones and load spikes.

If you're working with agentic systems, tool integrations must scale with them, especially if you're using a framework like the Model Context Protocol (MCP) to connect to enterprise tools. Monitor agent-triggered actions for error spikes and API failures.

Figure 5.11 outlines the core scaling considerations. Scale incrementally: expand regionally, test capacity thresholds, and don't overcommit to concurrency without proof.

Scaling Area	What to Consider
Resource Optimization	Are you managing compute and token costs efficiently?
Capacity Expansion	Can the system handle more users or requests?
Performance Stability	Does the system remain fast and stable under load?
Cost Management	Are costs sustainable as usage grows?
Regional Availability	Is the system accessible in new regions or markets?

Figure 5.11: Key scaling considerations for GenAI systems.

Expand capabilities with precision

Growth isn't just about serving more users; it's about serving them better. Use feedback from Stage 6 to guide expansion. Add languages, extend tool integrations, or introduce richer agent behaviors. If your users want integration with platforms like Salesforce or ServiceNow, consider leveraging MCP to streamline data flow across systems. Every new feature must be tested in isolation and in orchestration, especially for agentic use cases. Add opt-in approval flows before allowing autonomous actions in new domains. Use A/B testing to compare feature sets and detect regressions before a full rollout.

Optimize for efficiency, not just uptime

At scale, small inefficiencies become expensive. Cache common prompt-responses, implement model fallback logic to reduce latency, and monitor which features or tool calls drive cost spikes. Agentic systems, in particular, may waste tokens or over-trigger APIs, track their behavior using tools like LangGraph and LangSmith. Use observability data to simplify prompt chains, trim unnecessary model hops, and prioritize routes that deliver similar quality at lower cost. This is where "resilience" matures into "efficiency."

Maintain trust at scale

As your system grows, so do the risks. You're no longer watching every interaction, but your users assume you are. This is the moment to elevate your governance practices: version prompts, flag drift, audit agent decisions, and create dashboards to highlight high-risk interactions. Add interpretability where it's needed ("I chose this meeting slot because..."), and maintain transparency through uncertainty messages and undo options. Don't just fix forward—plan ahead. Prepare for model or tool deprecation, write playbooks for rollback, and automate regular audits of behavior and data freshness. Trust scales when stewardship scales.

Conclusion

You've now traveled the full arc of building a GenAI product from framing the right problem to designing the system, prototyping behavior, building the real product, launching it, and scaling it thoughtfully. But along the way, you've also seen that building GenAI systems isn't just about picking a model or writing prompts.

It's about much more:

- Understanding when and why to use GenAI
- Making the system trustworthy, usable, and empowering
- Designing for behaviors, not just features
- Handling complexity with discipline and empathy
- Thinking beyond launch, toward long-term evolution

Whether you're building a chatbot, a copilot, or a fully agentic system, you now have a practical playbook to go from an idea to a living, evolving product. But remember—in GenAI, reaching your first summit isn't the end of the journey. It's only the first season of many. AI technologies, user expectations, and competitive landscapes will shift faster than traditional roadmaps can predict. The teams that thrive will be the ones who treat every launch as a beginning, continuously learning, adapting, and navigating new terrain.

In today's world, the strongest builders are also the strongest learners. Later, in Chapter 9, we'll zoom in on the key decisions GenAI PMs face across that journey like build vs. buy, stack design, system autonomy, and post-launch optimization. If this chapter was about *how to build*, that chapter is about *how to decide*. But there's one more piece to the puzzle: *How will you know if it's working effectively and ethically?*

That's what the next chapter explores, how to evaluate your product's performance while ensuring it remains fair, trustworthy, and aligned with responsible AI principles.

Chapter 6: Measuring Success in GenAI — Metrics and Ethics

"Not everything that can be measured matters, and not everything that matters can be measured."
— William Bruce Cameron (often misattributed to Einstein)

When I first started working with GenAI, I focused relentlessly on metrics—accuracy rates, response times, user engagement. I remember building a chatbot that seemed flawless on paper: fast, relevant, and well-received by early users. It wasn't until someone flagged that it had inadvertently revealed a sensitive detail that I realized how misleading those surface metrics could be. The system had passed all the standard benchmarks—but it failed the one that mattered most: trust.

For those who are venturing into the world of GenAI, knowing how to measure success is essential. It's not just about what the system can do, it's about how it does it. This chapter explores the dual pillars of GenAI evaluation: metrics and ethics. In GenAI, metrics and ethics are two sides of the same coin, and you can't succeed by focusing on one and ignoring the other. This chapter explores how to evaluate GenAI systems through both lenses:

- Metrics that track how well the system performs.
- Ethics that ensure the system behaves responsibly and earns user trust.

Together, these elements provide a robust framework for evaluating GenAI, especially in the realm of agentic AI, where autonomous decision-making heightens the demand for precision, fairness, and accountability.

Traditional metrics like accuracy or uptime are still useful but they aren't enough. GenAI systems generate unpredictable, context-rich outputs: text, images, decisions. A chatbot might seem flawless until it spreads misinformation. An agent might automate a task but ignore a user's boundaries. With agentic AI, the stakes are even higher: these systems act on behalf of users, so missteps aren't just bugs—they're breaches of trust.

Building on the foundation laid in Chapter 5, where we explored the creation and scaling of GenAI products, this chapter tackles the pivotal question: "How do you know your system is working effectively and ethically?"

Here's what we'll cover:

- Technical evaluation methods to assess performance and reliability
- Ethical challenges and strategies to address them
- The interplay between metrics and ethics
- The broader societal impacts of GenAI systems

With practical examples, actionable steps, and insights into real-world tools, this chapter will help you design success criteria that go deeper than surface-level key performance indicators (KPIs). Whether you're developing a creative content generator or an autonomous AI agent, you'll walk away knowing how to measure what really matters not just performance, but trust. Let's explore what it really means to measure success in GenAI.

Why Metrics Matter in GenAI

Metrics are your system's reality check. They offer an objective lens into whether your GenAI product is delivering on its promise. Without them, it's easy to misjudge progress or worse, celebrate superficial success while deeper problems go unnoticed.

Metrics matter because they surface what's working and what isn't. They help answer questions like: Is our support agent really reducing time to resolution, or just deflecting users with generic answers? Are we seeing increased usage because the experience is valuable, or because users are struggling to get what they need?

They also unify your team around shared goals. Engineers, designers, and business leaders may approach problems from different angles but metrics provide a common language to define what "good" looks like, whether that means higher user satisfaction, operational savings, or improved decision accuracy.

Most importantly, metrics help you distinguish between activity and impact. High engagement doesn't always mean high value. What matters is whether your system is helpful, reliable, and aligned with user needs.

As GenAI systems scale, especially agentic ones that make decisions on behalf of users, metrics become even more essential. You're no longer just measuring output quality; you're tracking behaviors, decision pathways, and downstream consequences. Metrics aren't just performance indicators, they're trust signals.

Before we break down specific evaluation frameworks, let's begin with the most important one: your North Star Metric.

The North Star Metric

Before diving into the nitty-gritty of measuring your GenAI system, it's important to anchor your approach with a clear and unifying goal: the one metric that best reflects whether your AI is delivering real value. This is your *North Star Metric*, a single measure that captures the core impact of your product on users and the business. It serves as your compass, ensuring that the dozens of granular metrics you track—ranging from latency to satisfaction—ultimately roll up to something that matters.

Why is this important? GenAI systems often produce a flood of measurable outputs: tokens consumed, prompt lengths, model latency, usage frequency. Without North Star, it's easy to get distracted by metrics that optimize for surface-level engagement without delivering meaningful outcomes. A well-chosen North Star Metric aligns teams across engineering, product, design, and leadership. It clarifies trade-offs, focuses decision-making, and keeps everyone pointed toward long-term value.

Consider Grammarly, which has evolved its own approach to measuring impact. For years, it made intuitive sense to track how often users accepted AI suggestions—each one signaling a positive interaction. But in early 2025, Grammarly rolled out a more holistic metric: the Effective Communication Score (ECS). ECS reflects not just correctness, but also clarity, tone, inclusivity, and adherence to brand voice. It allows teams to measure the quality of communication outcomes—not just the frequency of individual actions (Grammarly 2025).

This shift enabled Grammarly to guide product strategy with more context. For instance, ECS helps enterprise customers identify communication breakdowns or monitor inclusivity at scale. It also provides a feedback loop: product teams can track which new features or model updates meaningfully improve ECS, and which might have unintended side effects.

The lesson for GenAI PMs is clear: your North Star Metric should reflect the outcome your AI exists to improve, not just the interaction it facilitates. If you're building a scheduling agent, it might be "meetings successfully scheduled without escalation." If it's an AI tutor, maybe it's "concept mastery over time." These types of metrics are not just easier to align around, they're truer to the product's purpose. And just like your product, your North Star Metric may evolve. The more you learn about how users interact with your system, the better you'll get at defining what "value" means.

How Do You Define Your North Star Metric?

A great North Star Metric doesn't emerge by accident—it's the result of thoughtful design. Here's a simple three-step process to define one that's actionable, aligned, and meaningful.

Step 1: Clarify your system's purpose

What core problem is your GenAI product designed to solve? And what lasting value should it create for users and the business? Your purpose should ground everything that follows.

Step 2: Identify the key user behavior

What user action (or small set of related actions) best signals that your product is delivering that value? In simpler systems, one behavior like completing a task or resolving an issue may be enough. For more complex products, a composite view may offer a more accurate reflection of success.

Step 3: Translate behavior into measurable metric

Define how you'll track the behavior over time. Your metric should be observable, outcome-driven, and scalable, one that surfaces user value clearly and guides decisions as your product evolves. The right metric not only reflects current performance but also helps prioritize future improvements.

This approach works across a wide range of GenAI products. For instance, a coding copilot might use "lines of code accepted by developers" to reflect its core value—speed and usefulness. A meeting-scheduling agent might track "meetings scheduled without human correction" to reflect autonomy and time saved.

Whatever your system does, aim to identify the one metric that reflects the outcome it exists to drive. That's the lens through which every other metric should be interpreted. But what happens when you don't define this clearly or pick the wrong one?

Pitfalls of a Missing or Misaligned North Star Metric

In April 2025, Anysphere's Cursor—a breakout AI coding assistant—faced a viral backlash when its customer support bot hallucinated a non-existent login policy. Users were unexpectedly logged out across devices, and the AI, posing as a human named "Sam," confidently explained the disruption as part of a new policy. But there was no such policy. The explanation had been entirely fabricated by the system. The issue wasn't discovered internally, it surfaced only after frustrated users raised the alarm publicly on Hacker News and Reddit, leading to subscription cancellations and reputational damage (Goldman 2025).

Cursor never disclosed what internal metrics guided its support automation. But that's the point: without clearly defined, user-centered metrics—such as resolution accuracy, trust alignment, or escalation clarity—AI systems can confidently deliver the wrong outcomes at scale. This failure wasn't just technical. It was systemic. The AI wasn't optimized around the values that mattered most to customers, like transparency and reliability.

For GenAI product teams, the lesson is clear: if you don't explicitly define success in terms of real-world value and user trust, your system may optimize for the wrong things like speed or response volume while silently undermining the very credibility it depends on. In a world where agents can act autonomously, a misaligned or missing North Star Metric isn't just a missed opportunity. It's a risk vector.

What Metrics to Measure for Different GenAI Systems

GenAI systems aren't one-size-fits-all, they vary in purpose, architecture, and user interaction patterns. From prompt-only LLMs to fully autonomous multi-agent systems, each has its own definition of success. The right North Star Metric should align with what your system is fundamentally built to do—and how users experience that value.

For instance, a retrieval-augmented generation (RAG) system must not only generate fluent responses but retrieve accurate supporting evidence. A copilot helps users complete tasks faster; a multi-agent system must coordinate several components without breaking the flow. Measuring success requires focusing on outcomes, not just activity.

Figure 6.1 summarizes the core objective, user-centered success signal, and a representative North Star Metric for each major GenAI system type. This table helps you choose metrics that reflect the real-world value your product is meant to deliver—not just what it outputs, but how well it works in context.

GenAI System	Core Objective	User-Centered Success Signal	Representative North Star Metric
LLMs (Prompt-only)	Generate accurate responses	Users get correct and helpful answers	Helpful Answer Rate
RAG	Retrieve and generate relevant answers	Users receive well-supported, context-aware outputs	Answer Relevance Score
Tools/Function Calling	Execute functions correctly	Functions complete as expected	Successful Function Call Rate
Copilots	Assist with task-specific productivity	Users finish tasks faster and with less friction	Task Completion with AI Assistance
Agentic AI	Perform autonomous tasks reliably	Tasks are done right without human correction	Autonomous Task Success Rate
Agentic RAG	Retrieve and act autonomously	Retrieved data powers accurate task execution	Retrieved Task Completion Rate
Multi-Agent Systems	Coordinate agents to complete workflows	Agents collaborate smoothly to meet user goals	Multi-Agent Task Success Rate

Figure 6.1: Objectives and success metrics for different GenAI systems

With your North Star defined, the next step is to measure how your system performs across dimensions that matter both technically and ethically. That's where the four-dimension evaluation framework comes in.

The Four-Dimension Evaluation Framework

Now that your North Star Metric is defined, the next step is knowing how to measure progress across all aspects of your GenAI system. This is where the four-dimension evaluation framework comes in—covering user experience, model performance, system reliability, and business outcomes. Adapted from Google's internal methods, it provides a practical, structured way for GenAI product managers to evaluate success in real-world settings. This section introduces key metrics for each dimension, including how they apply to more complex systems like agents. For more use case–specific metrics, see Appendix C.

Next, we'll go over how the four dimensions work, along with key metrics to track for each.

User Dimension: End-User Experience

The user dimension asks: *Are people using your GenAI system—and is it helping them do what they came to do? And most importantly—do they trust it enough to keep using it?* It captures how your AI system feels in the hands of real users.

Start with **adoption rate**—how many users are actively engaging with the system—and *task completion rate*—how often they successfully finish what they set out to do. You can also track **frequency of use** to understand how often the system fits into users' routines. These three signals alone can reveal a lot: a system with low adoption might not be discoverable or useful; a system with high adoption but low task completion might be confusing or ineffective; and infrequent use could suggest it's not sticky or essential.

Beyond usage, look for signs that your system is delivering meaningful value. Are users satisfied with the experience *(satisfaction score)*? Whether you collect feedback through thumbs-up ratings, post-task surveys, or open-text responses, satisfaction tells you not just what users did, but how they felt doing it. And in time-sensitive workflows, measuring how much time the AI saves *(time efficiency)* can offer a tangible indicator of its value.

For agentic systems that operate with more autonomy, additional metrics come into play. **Autonomy effectiveness** reflects how often the agent can complete tasks without human help. **Intent resolution** measures whether it correctly interprets what users are asking, especially when prompts are vague or open-ended. And user control captures whether people can steer the agent when needed, override decisions, or set boundaries.

Together, these metrics help you understand how user experience contributes to your North Star. For instance, if your North Star is "issues resolved without human help," and resolution rates are high, but user satisfaction is low, you may be hitting the metric while still eroding trust. The system might be efficient—but not empathetic.

Model Dimension: Evaluating AI Output Quality

This dimension focuses on the quality of your AI's responses—are they accurate, complete, and aligned with what the user asked? Here, you're not measuring how users feel, but whether the model's behavior meets your quality bar.

For prompt-based systems, this often comes down to **response accuracy** (is the answer correct?), **groundedness** (does it stick to source data?), and **coherence** (is it logically structured?). In other words, does the AI give clear, truthful, and well-formed answers?

As systems get more complex, additional model-driven signals become important. **Response completeness** measures whether the model fully answers the user's intent—if a user asks for a meeting to be scheduled, does the system provide time, location, and invitees, or just one of those things? **Task adherence** checks whether the model sticks to its intended role for instance, an assistant that's asked to draft an email but starts answering unrelated questions, that's a failure of adherence, even if the response itself is fluent.

And **decision consistency** reflects how reliably the model applies its logic across similar prompts, for example, generating the same summary format each time it's asked to summarize a report.

Ultimately, this dimension helps you answer a foundational question: Is the model doing what it's supposed to do, the way it's supposed to do it?

But even technically strong outputs can still cause harm if they're biased, misleading, or unsafe. That's why later in the chapter we would cover how to evaluate your AI system's ethical behavior especially in sensitive or high-impact use cases.

System Dimension: Evaluating Performance and Coordination

Not every behavior in a GenAI system comes from the model. Some issues arise downstream when the model hands off a task to tools, memory, or other system components. That's what this dimension captures: how well the entire system performs, especially when multiple moving parts are involved.

Sometimes, the model output is perfectly sound, but the tool fails, memory returns bad context, or two agents generate conflicting actions. These aren't failures of reasoning—they're breakdowns in orchestration, infrastructure, or coordination.

So here, you're asking: Does the system do what it's supposed to—reliably, responsively, and at scale—when the model interacts with the rest of the stack?

To evaluate that, look at signals in two broad categories: how well the system performs under load, and how well it coordinates across components.

System performance

Start by evaluating the raw performance of your infrastructure. How fast is your system when it matters most? *Latency* captures the total time a user waits—from submitting a prompt to receiving a response. It includes not just model inference, but also tool invocation, context lookup, and output streaming. High latency can break the flow of user interaction. *Uptime* measures how reliably your system stays available. Even occasional outages can disrupt usage patterns and erode trust. *Throughput* reflects your system's ability to process high volumes of data—whether measured in tokens per second or GPU utilization under load. And behind it all, *cost efficiency* determines whether the system is financially viable at scale—not just the model call, but every step in the task execution chain.

Together, these metrics reveal whether your system can perform under pressure—without introducing lag, failure, or runaway costs.

System coordination and integration

Beyond raw speed, GenAI systems—especially agentic or multi-agent ones—must coordinate well. *Tool call success* tracks whether the tools selected by the model actually execute properly; a failed API call or timeout is no less of a failure just because the model chose the right tool. *Action completion rate* tells you whether the entire workflow succeeded—whether it's a multi-step reasoning loop, a series of chained tools, or a coordinated agent pipeline.

In multi-agent setups, *behavioral stability* becomes critical: do agents behave consistently when presented with similar inputs, or do they shift unpredictably? And *conflict rate* helps you detect when agents interfere with one another, producing contradictory results or duplicative actions due to poor orchestration logic.

Together, these signals expose the hidden cracks in your system's architecture beyond just model quality. They help ensure your GenAI system performs under pressure and behaves as a cohesive unit.

Business Dimension: Connecting AI to Business Outcomes

Building a GenAI system isn't just about helping users, it's about delivering measurable value to your organization. This dimension reframes system performance through the lens of business stakeholders: *Is this investment paying off? Should we scale it further?*

As a product manager, you're not just evaluating functionality, you're articulating strategic impact. From revenue and retention to time-to-value and risk posture, these metrics help translate your AI efforts into business outcomes that executives care about.

Start with the **financial levers**. *Revenue growth* reflects how much income your system generates, whether through subscriptions, usage-based pricing, upsells, or one-time purchases. But top-line growth only matters if it's sustainable. That's where *profit margin* comes in: how much of that revenue remains after infrastructure, development, and support costs are accounted for. Meanwhile, *customer acquisition cost (CAC)* reveals how efficiently you're bringing in new users or accounts—whether via ad spend, demos, or outreach.

Retention is just as critical. A strong *revenue retention rate*, the percentage of recurring income you retain from existing users or customers, signals long-term product-market fit. And when users upgrade their plan, buy new features, or increase their usage over time, that's *contract expansion* (in B2B) or *monetization growth* (in B2C), clear signs that your product is delivering increasing value.

Next, consider how your product is performing in the market. Metrics like *customer acquisition rate* and *sales conversion rate* track your success in turning interest into adoption whether through trials, marketing funnels, or partnerships. A low *churn rate* means customers are sticking around. And if your *market share* in a category like "AI for writing," "AI for developers," or "AI for customer support" is expanding, you're gaining ground against competitors.

Then there's time-to-impact. *Time to value (TTV)* measures how quickly users (individuals or organizations) begin seeing benefits like time saved, cost reduced, or productivity boosted. Short TTV can strengthen word-of-mouth, increase renewals, and accelerate growth. So can a strong *Net Promoter Score (NPS)*, especially from power users, champions, or decision-makers who influence others.

Finally, you'll want to stay ahead of risk and compliance. *Compliance adherence* ensures your system meets regulatory standards like GDPR, HIPAA, or CCPA, especially if your product handles sensitive data or operates across regions. And if something goes wrong, *incident rate* and *mitigation effectiveness* measure how well your team detects, addresses, and prevents future issues preserving customer trust and protecting revenue.

Note: Not all metrics apply equally to every GenAI product. The right ones depend on whether you're selling to consumers or businesses, and where your product is in its lifecycle. But as a PM, it's your job to connect your AI system to measurable business outcomes—and choose the metrics that matter most.

This framework ensures a balanced evaluation. Too often, I've seen product managers over-index on a single dimension like obsessing over system latency while overlooking user trust, or fine-tuning model accuracy without considering business impact. The four-dimension view helps you step back and assess your GenAI system holistically across users, models, systems, and business outcomes.

In the next section, we'll explore how to put this framework into action showing how each dimension contributes to your North Star Metric.

Applying the Four-Dimension Evaluation Framework

The four-dimension framework isn't just a checklist, it's a way to connect day-to-day metrics back to your North Star Metric, making it easier to spot what's working, what's broken, and what needs investment.

Let's take customer support, one of the first business functions meaningfully disrupted by GenAI. Imagine you're building an AI-powered support assistant, and your North Star is: "Issues resolved without human help."

This single outcome relies on multiple parts of your GenAI system working in harmony and each dimension helps you evaluate a different link in that chain.

The *user dimension* tells you whether the experience feels successful from the customer's perspective. Are people getting their problems solved? Are they satisfied with how it felt? If completion or satisfaction scores are low, your resolution rate might be hitting the target while still eroding trust suggesting a need for better UX, tone, or control.

The *model dimension* looks beneath the surface to ask: *Are the responses any good?* This is where you check if your assistant's answers are accurate, complete, and well-structured. If customers are bailing because the AI gives inconsistent or vague answers, the fix might lie in prompt tuning, grounding, or retraining not UX.

The *system dimension* reveals whether everything functions under real-world conditions. Does the system respond fast enough? Do tool calls and memory access succeed reliably? A high abandonment rate might not be about the model or UX at all, it could be the result of backend delays or failed orchestration.

And finally, the *business dimension* reframes everything in terms of organizational value. Even if your assistant resolves issues well, does it reduce human workload or support cost? Are those benefits translating into contract renewals or upsells? If not, you might be building a technically sound experience that's underperforming strategically.

By looking across all four dimensions, you're not just debugging your GenAI system, you're deconstructing your North Star, piece by piece, and mapping each part back to the metrics that influence it. That's how product managers turn insight into impact. In the next section, we'll explore how to layer ethical evaluation into your metrics ensuring your GenAI system is not just effective, but trustworthy.

Ethical Considerations in GenAI

Now that you've defined your North Star metric and explored how to measure progress across user, model, system, and business dimensions, it's time to tackle something just as vital: ensuring your GenAI system does the right thing. Metrics like task completion and cost savings show how well your system performs but they don't tell you if it's behaving responsibly.

I learned this the hard way when I once launched a GenAI assistant—users loved its speed, but my stomach sank when I realized it had shared a user's personal detail in a response. Great numbers mean nothing if trust takes a nosedive.

Users expect your AI to be fair, transparent, and trustworthy, especially in sensitive domains like healthcare, finance or hiring. Ethics isn't a nice-to-have; it's a must, particularly for agentic AI systems that act on behalf of users, like scheduling meetings or managing workflows, where a biased decision or privacy breach can cause real harm. This isn't just about reputation. It's about responsibility.

Let's explore the key ethical challenges GenAI systems face, how to measure them, and how to incorporate them into your North Star metric.

Common Ethical Challenges in GenAI

Ethical risks are baked into the way GenAI works. These systems generate content, simulate reasoning, and make decisions based on massive datasets—many of which contain historical bias, inaccuracies, or sensitive information. As a result, product managers must confront four recurring challenges that can quietly undermine even the most impressive systems. Figure 6.2 summarizes the main ethical challenges you need to watch out for, along with their impacts and real-world examples.

Ethical Challenge	Impact on Your System	Real-World Example
Bias and Fairness	Alienates users, causes harm, erodes trust	ChatGPT gender bias (2023)
Transparency	Erodes trust, frustrates users	Credit scoring tool opacity (2024)
Privacy and Data Security	Destroys trust, risks legal issues	Samsung-ChatGPT data leak (2023)
Misinformation and Harm	Misleads users, damages reputation, causes harm	Google Bard misinformation (2023)

Figure 6.2: Ethical challenges in GenAI systems

The first is *bias and fairness*. If your model is trained on skewed data, it will mirror those biases in its outputs. In 2023, a study revealed that ChatGPT frequently associated leadership roles with men and caregiving roles with women—subtle but damaging patterns that can alienate users and reinforce harmful stereotypes (The Guardian 2023).

Next is *transparency*. When GenAI systems are used in high-stakes domains like finance or healthcare, users want to know how decisions are made. In 2024, users of a credit scoring tool powered by GenAI were frustrated after being denied loans without any explanation. The lack of clarity didn't just cause confusion, it led to public backlash and a loss of trust (Forbes 2024).

Then there's *privacy*, a critical concern as GenAI often processes sensitive user data. In one high-profile case from 2023, a Samsung employee accidentally pasted proprietary source code into ChatGPT. That code was then retained by the model, raising serious concerns about intellectual property and corporate confidentiality (Vincent 2023). Finally, there's the risk of *misinformation and harmful outputs*. GenAI systems can confidently generate convincing but false information. In 2023, Google's Bard wrongly claimed that the James Webb Space Telescope had taken the first image of an exoplanet—a factual error that quickly spread and damaged the system's credibility (Klee 2023.

Each of these risks—bias, opacity, leakage, and misinformation—has the potential to erode user trust, damage your product's reputation, and cause real harm. And as your system gains more autonomy like taking actions, making decisions, or triggering workflows, these risks only grow more consequential.

Integrating and Measuring Ethics as Metrics

Ethics can't be a side conversation. It needs to be built into your evaluation framework, right alongside four-dimension metrics to ensure your system operates fairly, transparently, and responsibly. Figure 6.3 summarizes the four core ethical principles, showing what to track, why it matters, and how it applies to general and agentic AI.

Ethical Principle	What to Measure	Why It Matters	Example	Agentic Systems Considerations
Fairness	Demographic parity, Equal error rate, User flag rates by cohort	Prevents bias that can upset users	Equal loan approval rates for all demographics	Log interactions when affecting sensitive decisions
Transparency	Explainability scores, Reason visibility, Source traceability	Builds trust by showing users how decisions are made	95% of chatbot responses include an explanation	Track clicks on "See why" for explanations
Privacy	PII exposure rate, Anonymization coverage, Prompt leakage checks	Protects user data and avoids legal issues	Percentage of responses with personal details	Scrub memory before acting to avoid data leaks
Accountability	Feedback resolution rate, Escalation triggers, Model update responsiveness	Ensures quick fixes to maintain trust	Percentage of user-reported issues resolved	Allow, review, and act on overrides

Figure 6.3: Key ethical metrics for GenAI systems

Here's what each principle means in practice:

Fairness

Are outcomes consistent for all users? Measure parity in false positives/negatives across demographics like the same rate of incorrect rejections (false negatives) for men and women. Unequal rates might mean bias, such as flagging one group more often.

Transparency

Can users understand your AI's decisions? Track reason visibility (does the user see why something happened?) and source traceability (can they verify the answer's source?). For example, if a GenAI assistant denies a loan, does it explain why?

Privacy

Does the AI protect sensitive information? Track metrics like *PII (personally identifiable information) exposure rate* and *prompt leakage frequency.* In high-risk domains, a single privacy breach can mean lawsuits or lost customer trust that never recovers.

Accountability

Mistakes are inevitable. What matters is how quickly you detect and fix them. Measure feedback resolution rates (how many flagged issues get fixed?) and escalation triggers (how often the AI calls for human help). Also measure model update responsiveness—when something goes wrong, how long until it's addressed in production?

These metrics act as an early warning system, helping you spot issues like bias or privacy risks before they harm trust, especially in agentic systems where actions can have bigger impacts. By weaving these metrics into your system's telemetry and product reviews, you build a system that's not only functional but principled.

Think of these not as legal checkboxes, but as trust signals. They help you build systems that adapt ethically as they scale, what I call ethical agility. In the next section, we'll explore how to balance these ethical metrics with other metrics, ensuring your GenAI system succeeds on both fronts.

Balancing Ethics and Metrics

By now, you've built a full metrics stack—tracking how your GenAI system performs across user experience, model quality, system reliability, and business outcomes. You've also layered in ethical metrics to ensure fairness, transparency, privacy, and accountability.

But what happens when these goals conflict? You may find yourself optimizing for cost or speed, only to see transparency drop. Or pushing for higher automation, while unintentionally increasing bias or reducing user control. These are not edge cases, they're common trade-offs in GenAI development. This section offers practical guidance for navigating those tensions. Not every decision will optimize every metric. But by approaching trade-offs intentionally, you can ensure that your system stays aligned with both your product goals and your ethical commitments.

Navigating Trade-Offs: Real-World Strategies

Balancing ethical and non-ethical metrics doesn't mean sacrificing innovation, it means designing with awareness and intent. When trade-offs arise, your role is to make intentional choices that preserve trust, even if it adds some friction.

This balancing act becomes far easier when ethics are embedded in your North Star from the beginning, not added at the end as damage control. Here are four practical strategies to guide those moments:

Make reversibility part of your design.

GenAI systems evolve quickly—and so do user expectations. That's why reversibility matters. When GitHub Copilot introduced an "undo" and "recall previous suggestions" feature, it gave developers more control over the assistant's output. This wasn't just a usability tweak, it was a trust-building mechanism. By allowing developers to roll back AI suggestions, GitHub created a version-controlled feedback loop that empowered users without slowing them down. Designing reversibility—whether through opt-in behaviors, editable prompts, or review toggles—ensures that experimentation doesn't become a one-way street.

Prioritize clarity over smoothness.

Natural-sounding responses may impress users at first, but clarity builds confidence over time. When Slack rolled out its AI summarization tool in 2024, users praised its helpfulness but flagged that the summaries sometimes felt vague. In response, Slack added time-stamped references and rationale cues like "This was discussed on Monday's thread." It made the summaries a little more robotic but a lot more trustworthy. Especially when AI is making decisions or surfacing priorities, visible logic is often more valuable than linguistic polish.

Default to inclusive, neutral options.

Inclusivity isn't just about fairness; it's a pathway to broader engagement. Grammarly learned this while redesigning its tone suggestions in 2024. Originally, the assistant often nudged users toward overly casual or assertive styles—unintentionally reinforcing dominant cultural norms. After feedback from diverse user groups, Grammarly retrained the model on a wider range of writing styles and shifted the default to a neutral tone. Other tones remained available, but they were now opt-in and clearly labeled. This adjustment didn't just improve fairness, it increased usage among underrepresented groups and built stronger affinity.

Design for human handoff.

Autonomy can be powerful—but in sensitive domains, escalation paths are essential. In early 2025, Cerence launched its xUI in-car agent platform with a hybrid LLM architecture. In routine tasks, the system handled navigation and commands fluidly. But when safety-critical decisions arose—like ambiguous routes or complex merges—the agent deferred to the driver, prompting: "Driver input required for safety." This wasn't a fallback. It was an intentional design constraint—embedding human judgment where it mattered most. That kind of proactive handoff is what keeps autonomy aligned with accountability.

Final thought

Ethical success in GenAI isn't a fixed outcome, it's a continuous loop. Observe your metrics. Analyze the trade-offs. Improve your system. That's how you move from good intentions to responsible product leadership. You observe your metrics—not just usage, latency, and retention, but fairness, explainability, and autonomy. You analyze the trade-offs—recognizing when progress in one metric quietly erodes another. And then you improve—by refining your models, prompts, defaults, or guardrails.

The best GenAI teams don't wait for crises. They design with foresight and iterate with care. Because in the long run, trust isn't just a value, it's the most defensible product moat you have.

Tools to Measure What Matters

Knowing *what* to measure isn't enough, you need the right tools to track, learn, and act. In GenAI, you're measuring hard-to-see things like bias, hallucinations, or user trust—especially in agentic systems. This section shows how to choose and apply tools that match your system's context and maturity.

Choosing the Right Tool for Your GenAI Product

Choosing a GenAI tool isn't about chasing features, it's about finding the right match for your system's goals, maturity, and risk. What works for a prototype LLM might fail in a production-grade agentic system. Instead of starting with a product comparison, start by asking: What are we building? Where are we in the lifecycle? What could go wrong?

Figure 6.4 outlines five core factors to guide your selection. Think of these not as a checklist, but as a framing exercise:

Factor	What to Consider
System Type	Are you building an LLM app, chatbot, RAG system, or agentic AI? The more autonomous it is, the more you'll need behavioral observability and trace replay.
Product Stage	At prototype: favor fast feedback and explainability. At scale: you'll need tools that offer transparency, fairness tracking, and drift monitoring.
Risk Profile	What could go wrong? If bias or safety is a risk, choose tools with fairness scores or PII logging. Ethical awareness is part of risk—not separate from it.
AI Tech Stack	What platforms and infra do you use (e.g., Azure, Google, AWS)? Tools must integrate easily with your workflows and support your compliance needs.
Compliance and Governance	Does your org require audit logs, GDPR/CCPA compliance, or explainability-by-default? These policies can restrict what tools are approved.

Figure 6.4: Key factors for choosing the right tools

First, consider your *system type*. Are you building a simple chatbot or an agentic system that takes action on its own? The more autonomous your system, the more you'll need tools that provide behavioral traceability—tools that let you track not just what the model said, but what it decided to do and why.

Next, think about your *product stage*. In early development, speed and explainability are key. You'll want tools that help you iterate quickly and understand failures. But as you move toward scale, the focus shifts. Now, you'll need to monitor drift over time, ensure fairness doesn't degrade, and track reliability metrics like uptime and system latency.

Your *risk profile* should also shape your toolset. If your system handles sensitive data or makes high-impact decisions, ethics isn't a side consideration, it's central. Prioritize tools that can detect bias, monitor safety violations, and log decisions in ways that support auditability.

Tool compatibility also depends on your *AI tech stack*. Whether you're working on Azure, Google Cloud, AWS, or a custom environment, the tools you choose need to plug into your infrastructure smoothly supporting telemetry pipelines, tracing systems, and retraining workflows.

And finally, factor in *compliance and governance*. Some organizations, especially those in healthcare, finance, or public services have strict rules around data handling, audit logging, and explainability. Your tooling choices may be limited by these requirements, so it's important to understand what's approved early on, rather than getting blocked late in the review process.

By grounding your tool selection in these five factors, you ensure your observability stack isn't just powerful, it's practical and aligned with your product's needs. In the next section, we'll map specific tools to your product's lifecycle, so you can prioritize the right insights at the right time.

Tools Across Your AI's Lifecycle

As your GenAI system matures, so should your tools. Each stage of development—*prototype*, *launch*, and *scale*—requires a different kind of visibility.

Early on, it's about learning fast. Later, it's about operating reliably and ethically at scale. Figure 6.5 lists what to track at each stage and tools to consider.

Stage	What to Track	Tools to Consider
Prototype	Prompt quality, speed, hallucinations	PromptLayer, Azure AI Foundry
Launch	User satisfaction, output safety, undo ability	PromptLayer, Azure Monitor, OpenAI Moderation API
Scale	System reliability, fairness over time, cost, safety, data drift, resource use	Traceloop, LangSmith, Weights & Biases, Guardrails AI

Figure 6.5: Tools across the AI lifecycle

During the *prototype* phase, you need fast feedback loops. Tools that help evaluate prompt behavior, surface hallucinations, and monitor generation speed are key. *Prompt evaluation platforms* like **PromptLayer**, **LangChain Eval**, or custom logging scripts help you refine interactions before real users are involved.

At *launch*, you shift focus to safety and user experience. You'll want tools that can monitor for harmful content, trace model decisions, and give users the ability to undo actions. Options like **moderation APIs**, **user feedback pipelines**, and **observability dashboards** (such as those built on **OpenTelemetry**) can help here. If you're using a specific cloud platform like Azure, Google Cloud, or AWS leverage their built-in monitoring and alerting capabilities for production stability.

At *scale*, your priorities expand: you need to track *system reliability*, *fairness over time*, *data drift*, and *resource efficiency*. Tools like **LangSmith**, **Weights & Biases**, **Traceloop**, and **Guardrails AI** support traceability, fairness evaluation, and ethical guardrails regardless of your cloud provider. These tools are often cloud-agnostic or offer integrations across environments.

Across all stages, aim for tool coverage across your *four-dimension framework*: **PromptLayer** or user survey tools capture the *user dimension*, **Weights & Biases** or **MLflow** help assess *model quality*, platform-native or OpenTelemetry-based dashboards cover *system metrics*, and billing + usage analytics track *business impact*. Matching tools to your stage and stack ensures you're not just building fast, you're building with insight.

Observing Agentic AI: Tools That Track What It *Does*

Agentic systems don't just generate outputs, they take action. That makes observability more complex and more essential. You're not just evaluating what the model says, you're evaluating what it *does*: which tools it calls, which decisions it makes, and how reliably it follows rules. To do this well, you need tools built for behavioral tracing, not just output scoring. These tools should let you:

- Log decision steps (what did the agent choose to do, and when?)
- Replay behavior to diagnose issues when things go wrong
- Track user overrides to see how often humans intervene
- Enforce boundaries (like fallback logic or escalation triggers)

Tool	What It Helps You Measure
Traceloop	Logs and replays AI actions, ideal for checking Decision Consistency and Behavioral Stability.
LangSmith	Tracks AI decisions and actions, helping you measure Autonomy Effectiveness and User Control Rate.
Guardrails AI	Sets boundaries, blocks risky actions, and ensures Ethical Alignment by keeping actions safe and fair.
Azure AI Foundry	Measures a range of metrics like Intent Resolution (how well the AI understands user requests), Tool Call Accuracy (correct tool usage), and Code Vulnerability (security risks in generated code), plus safety checks with red teaming to keep agents trustworthy.

Figure 6.6: Agentic tool highlights

As of 2025, several tools are designed with this kind of *agentic observability* in mind. Traceloop offers behavior logging and replay. LangSmith helps assess *decision consistency* and *autonomy effectiveness*. Guardrails AI enforces policy boundaries and blocks unsafe actions. And Azure AI Foundry *(if you are on Azure Stack)* helps track everything from *intent resolution* and *tool call accuracy* to *code safety*. Figure 6.6 summarizes what each of these tools helps you measure.

If you're deploying agentic AI in production, tools like these aren't optional, they're part of keeping your system accountable, explainable, and under control.

These tools don't just log outputs, they help you trace decisions, enforce guardrails, and iterate safely. Whether your agent is scheduling meetings or prioritizing emails, its actions should be *observable, reviewable,* and aligned with your system's intent. If an agent edits a doc or deletes data, you should know *what* it did, *why,* and *how to fix it* when things go wrong.

OpenTelemetry: Why It Matters

As observability standards evolve, *OpenTelemetry* is emerging as the backbone for ethical, scalable monitoring in GenAI. Many modern tools like Traceloop, Azure Monitor, and Weights & Biases now integrate with it, making it easier to track how your system behaves at scale.

Here's why it matters:

- It lets you trace requests *end-to-end*, from user input to model output to tool execution, so you can reconstruct what happened and why.
- It supports agentic observability, helping you track decisions and actions over time, not just outputs.
- It standardizes how you surface ethical risks like fairness violations or unsafe tool usage so you can detect problems early and respond confidently.

OpenTelemetry doesn't measure ethical outcomes directly, but it provides the *technical foundation* to do so at scale. If you're working in regulated industries or deploying multi-agent systems, its logging and traceability are especially valuable for compliance.

As a GenAI PM, adopting tools that support *OpenTelemetry* ensures you're building not just functionality, but visibility and responsibility into your system's DNA.

Tools Reflect What You Prioritize

The tools you use don't just measure what's happening, they shape how your team builds. Your observability stack sends a clear message: *This is what we care about. This is what we watch.*

When chosen intentionally, the right tools become part of your product muscle. They guide how you define success, catch problems early, and evolve responsibly, especially as your system scales in complexity and reach.

Here's how to align your tool choices with your product values:

Anchor to what matters most: Start with your North Star metric and ethical KPIs especially those tied to trust, like fairness, transparency, and safety.

Cover your four dimensions: Choose tools that offer visibility across user experience, model quality, system performance, and business impact.

Grow intentionally: As your system matures, expand from lightweight logging and prompt evaluation to deep traceability and bias detection.

Track not just output, but behavior: For agentic systems, the real impact isn't what the model says, it's what it *does*.

As one senior PM at a Fortune 500 AI company put it: *"If you don't track trust, you can't fix it. Our observability stack isn't just about uptime, it's about ethics."*

By selecting tools that reflect both your values and your system's stage, you're not just tracking metrics, you're reinforcing *how* your team thinks, builds, and leads.

Looking Ahead: The Future of Measuring Success in GenAI

GenAI is evolving and so are the metrics that define success. As systems grow more autonomous and regulations tighten, the next generation of tools and standards will demand more from product managers than ever before.

Smarter tools are coming. In 2025, AI-driven tools already help track fairness and hallucinations. But soon, they'll predict risks before they happen flagging bias patterns, tracing agent behavior in real time, and enabling proactive debugging in complex multi-agent systems.

So are stricter rules. The EU AI Act now mandates metrics for clarity and safety in high-risk domains like healthcare and finance. By 2026, metrics like *explanation coverage* or *auditability rate* may become legal requirements, not just ethical aspirations.

Agentic systems raise the bar further. With AI agents coordinating, reasoning, and acting, metrics will go beyond accuracy into *behavioral consistency*, *collaboration fidelity*, and *value alignment*.

These shifts build on everything you've explored in this chapter: not just how to track what your AI does, but how to ensure it behaves as intended. But metrics alone don't define value. In the next chapter, we turn to what truly defines value: business models. You'll explore how GenAI products, especially those powered by agents, are being monetized, sustained, and scaled in the real world.

Chapter 7: Emerging Business Models in GenAI (Including Agentic AI)

By now, you've explored what it takes to design a GenAI system that works—technically, ethically, and in the real world. You've seen how AI can reason, summarize, plan, and respond. But here's the truth facing product teams today: *The harder challenge isn't building something impressive. It's building something that lasts.*

Adoption has surged. As we explored earlier in *Chapter 2*, companies have moved quickly to integrate GenAI across functions. But impact is a different story. A McKinsey report found that while 71% of companies now use GenAI in at least one business function, over 80% have seen no measurable impact on earnings (McKinsey & Company 2025). The gap isn't in capability, it's in capture.

The problem isn't that GenAI systems don't work. It's that too few have business models that do. The model works. The demo lands. But what's the business?

Too few GenAI products are built on business models that scale. Fewer than 20% track success through clear KPIs. Many lack pricing strategies that align with what their AI actually does. And as GenAI moves from novelty to necessity, the window for figuring this out is closing fast.

This chapter is about helping you bridge that gap. We'll explore the economics of GenAI—how companies create, deliver, and capture value across the stack—and how you, as a PM, can design business models that support intelligent systems, not just features.

New Technology, New Economics

GenAI isn't just a better tool. It's a new foundation for software, services, and decisions. It can generate, summarize, synthesize and increasingly, it can reason, plan, and act. And that changes the game.

- Value isn't just in features—it's in intelligence.
- Performance is harder to price—what's the value of a good judgment?
- User expectations have shifted—they want speed, utility, and ethics.
- And success isn't just about adoption—it's about trusted impact.

As product builders, that means we're not just designing user flows, we're shaping value chains, economic incentives, and business logic that didn't exist five years ago. Traditional SaaS pricing was tied to seats or usage volume, but GenAI introduces deeper questions:

- Do we charge per query? Per action? Per completed task?
- How do you price judgment, or trust?
- What does "premium" mean when multiple models can produce similar outputs?
- And how do you make margins work when one user action might trigger ten model calls behind the scenes?

Agentic systems push this even further. When AI starts planning steps, invoking APIs, or making decisions for users, the cost and value structures become harder to predict and harder to contain. You're no longer selling static software. You're monetizing cognition. That means product teams must design not just user flows, but *economic logic*—aligning outcomes, incentives, and pricing with what the system actually does.

These shifts are reshaping the business logic of GenAI. In the next section, we'll break down five key trends that are redefining how GenAI companies create, capture, and sustain value.

Key Trends Reshaping GenAI Business Models as of 2025

As GenAI adoption accelerates, the economics of value creation and capture are being redefined. As of 2025, companies are no longer competing just on who has the most powerful model but also on how intelligently that model is integrated, orchestrated, and trusted.

Here are five trends shaping the next generation of GenAI business models:

Models Are Becoming Commoditized

As of 2025, many top-tier models—GPT-4 Turbo, Claude 3, Gemini 1.5 Pro, Mixtral, and Llama 3—perform comparably on general-purpose tasks like summarization, reasoning, and basic Q&A. This makes them functionally interchangeable in many use cases, especially when abstracted through orchestration frameworks like LangChain or Semantic Kernel.

At the same time, open-weight models like Llama, Mistral, and DeepSeek are improving rapidly. Combined with falling token prices for closed models, this is eroding the pricing power of proprietary APIs. Gartner estimates that over 70% of enterprises will evaluate or adopt open-source models for at least some workloads in 2025, further democratizing access. As model performance converges and switching costs drop, raw intelligence is no longer a sustainable moat. Model providers are being pushed to compete not just on accuracy, but on how easily their models can be integrated, customized, and scaled.

Some, like DeepSeek, are leveraging cost-efficient architectures to undercut incumbents on pricing. Others are moving up the stack—differentiating through enterprise-grade tooling, private deployments, orchestration APIs, and vertical fine-tunes tailored for domains like healthcare or legal.

Multi-model Strategies Are Becoming the Norm

No single model can do it all. Most enterprise-grade GenAI systems now combine multiple models, each optimized for a different task. A contract analysis tool, for example, might use one model to summarize a document, another to extract obligations, and a third to suggest revisions.

This shift is giving rise to orchestration platforms like LangChain, LangGraph, and Semantic Kernel that abstract model routing, memory, and tool invocation. But with greater flexibility comes added complexity. Each model adds cost, latency, and governance considerations, especially in agentic systems where a single user task might trigger multiple model calls. For GenAI PMs, this means you're not just picking a "best model", you're designing a system architecture where models work together. And every model you add is another node in your cost structure and compliance plan.

Agentic AI Is Shaping Outcome-Based Pricing

Traditional GenAI systems respond to prompts. Agentic AI goes further, it reasons, plans, and executes across steps, using tools and memory to complete complex tasks. That shift unlocks a new pricing logic. Instead of charging per token or per user, businesses are beginning to price by completed outcomes. Think: *automate ten support tickets, approve five invoices,* or *schedule three meetings.*

This reframes how GenAI products are monetized. You're no longer selling output; you're selling completed tasks. For product managers, that means defining value not in terms of usage, but in terms of impact. Outcome-based models also align better with enterprise ROI expectations, especially in regulated or mission-critical workflows.

Trust and Safety Are Becoming Business Model Levers

As GenAI systems expand into sensitive domains—healthcare, finance, legal—trust is no longer just a value—it's a feature people are willing to pay for. Enterprises now evaluate models not just on performance, but on privacy, explainability, and compliance with ethical standards. This has turned responsible AI into a monetization strategy. Providers like Anthropic and Cohere are emphasizing safety-aligned models, while others like Mistral offer on-prem or virtual private cloud (VPC) deployments to meet governance and data sovereignty needs.

An IBM study found that 65% of enterprises prioritize compliance-certified AI when making purchasing decisions, driving demand for products that offer auditability, alignment, and transparency (IBM 2024). This shift is also creating new revenue streams around governance tooling, compliance APIs, and AI ethics consulting. For GenAI PMs, this means trust isn't just about risk, it's about opportunity. Safety, privacy, and compliance can become differentiators that justify premium pricing, especially in regulated industries.

Specialization Is the New Differentiator

With general-purpose capabilities leveling out, GenAI companies are carving out new defensible positions by going deep—either into industries or user workflows. Many are fine-tuning base models for specialized use cases in healthcare, legal, finance, or education, where domain context can unlock accuracy and trust that generic models can't provide. Others are embedding GenAI directly into the tools people already use like GitHub Copilot for developers or Microsoft Copilot Studio for enterprise productivity, making the model invisible and indispensable.

Some open-weight providers like Meta and Mistral have adopted freemium strategies, offering models for free but charging for enterprise support, licensing, or usage beyond defined thresholds (e.g., Meta's 700M MAU cap). In all cases, the model isn't the product, the *experience and ecosystem* are. For GenAI PMs, this reinforces a key principle: your moat is less likely to be in the model itself and more likely to lie in how that model is fine-tuned, embedded, or orchestrated to deliver sustained value.

Together, these trends point to one truth: *GenAI is no longer just a feature—it's an economy. Business models are evolving from "access to intelligence" toward outcomes, orchestration, and trust. For product managers, this means success isn't just about the model you choose, it's about how you use it to build systems that deliver sustainable, defensible value.*

In the next section, we'll zoom out to see how these dynamics play out across the GenAI value chain and why your monetization strategy depends on where in the stack your product lives.

From Layers to Logic: Mapping Business Model Archetypes

Back in Chapter 2, we explored the evolving GenAI value chain—the major players shaping the ecosystem, from infrastructure providers and model labs to orchestration frameworks and application builders. But knowing *who* is involved is only half the story. The next question is: *How does each layer create, deliver, and capture value?*

This matters whether you're deep in the ecosystem—at a company building foundational capabilities like model APIs or GPU infrastructure—or on top of it, designing GenAI products that rely on those services. Business models are not just background mechanics, they shape what gets built, who can afford it, and how products scale.

Unlike traditional SaaS, GenAI economics are shaped by volatile usage, expensive inference, model licensing, and trust in probabilistic outputs. You're not just monetizing features, you're managing operational costs, regulatory constraints, and ecosystem dependencies.

This chapter bridges the value chain with long-term sustainability. It outlines five core business model archetypes, summarized in Figure 7.1, that reflect distinct strategies for capturing value across the GenAI landscape. Later, we'll explore how Agentic AI is giving rise to new models where systems don't just generate—but reason, plan, and act.

Archetype	How They Create Value	How Value Is Delivered	How Value Is Captured	Examples
Infra Providers	Power GenAI through compute, runtime, and managed hosting	Cloud infrastructure, managed runtimes, GPUs, APIs/SDKs	GPU/hour billing, MACC bundles, reserved capacity, CUDA software, inference endpoints	NVIDIA (incl. NIMs), Azure, AWS, Google Cloud
Foundation Model Providers	Deliver generalist or domain-specific intelligence via LLMs	APIs, fine-tuning services, hosted model access	Pay-per-use APIs, PTUs, tiered subscriptions, model licensing, managed fine-tuning	OpenAI, Anthropic, Microsoft, Meta, Mistral, Cohere, DeepSeek
Platforms & Marketplaces	Aggregate, customize, and distribute models or agents	Unified APIs, fine-tune hosting, model catalogs	Pay-per-use APIs, fine-tune hosting, inference endpoints, revenue share	Azure AI Foundry, Hugging Face, AWS Bedrock
ISVs / Vertical AI Builders	Embed GenAI into regulated or domain-specific workflows	Vertical apps, compliance-ready integrations	Vertical SaaS, task-based pricing, compliance add-ons, usage-based APIs	Saifr, Cognitiv+, SightMachine
AI-Native Apps & Ecosystem Hubs	Deliver GenAI to users via intelligent apps or productivity tools	Consumer apps, embedded assistants, productivity platforms	Freemium → Pro subscriptions, embedded monetization, in-app upgrades	GitHub Copilot, Microsoft Copilot Studio, Notion AI, TikTok Symphony, ChatGPT, Grok

Figure 7.1: Business model archetypes across the value chain.
© 2025 by the author of *Zero to GenAI Product Leader*.

While many companies focus their monetization at one dominant layer, others operate across archetypes. These hybrid strategies can unlock more value, but they also introduce complexity for GenAI PMs navigating pricing, integration, and product scope. So, let's start at the foundation with infrastructure providers, who define the economics of every GenAI interaction.

Archetype 1: AI Infrastructure Providers

Where GenAI Costs Begin and Margins Start to Shrink

Think of your GenAI product, whether a chatbot or an autonomous agent, as a car. It needs a powerful engine to run smoothly. That engine is the AI infrastructure layer: GPUs, compute clusters, optimized runtimes—all of which we explored in Chapter 2. This layer powers everything from inference to deployment.

But infrastructure doesn't just determine performance, it shapes your margins, cost structure, and scalability. In this archetype, we'll explore how infrastructure providers create, deliver, and capture value and why decisions made here ripple through the entire GenAI product lifecycle. Infrastructure isn't just a technical choice. It's a strategic one.

How value is created

AI infrastructure providers create value by turning raw compute into scalable, developer-ready platforms. They solve a foundational bottleneck: delivering the performance and reliability needed to train, fine-tune, and serve increasingly complex models and agentic systems—without teams needing to manage hardware themselves.

As GenAI evolves toward multi-step reasoning, real-time tool use, and persistent memory, compute needs are rising—even as models get leaner.

Infrastructure providers create this value through the following:

- AI-optimized hardware like Nvidia GPUs or custom silicon (e.g., Microsoft's Maia, AWS Trainium) designed to boost GenAI performance and lower costs.
- Cloud platforms (Azure, AWS, GCP) offer GPU clusters and elastic environments for both experimentation and production.
- Specialized GPU clouds (e.g., CoreWeave) tuned for GenAI, offering faster provisioning and cost-efficient access to premium hardware.
- Managed runtimes and developer services that abstract infrastructure complexity, allowing teams to scale without deep DevOps investment.

In short, they turn compute into a service letting product teams focus on delivering value, not managing hardware.

How value is delivered

Infrastructure providers deliver value by abstracting complexity. They turn hardware-intensive operations—racks of GPUs, cooling systems, cluster orchestration—into clean, developer-friendly services. As a GenAI PM, this means you don't need to configure a data center to scale your product. You need an API key, a runtime, and a credit card.

But not all infrastructure is delivered the same way. Delivery strategies vary, depending on performance requirements, cost sensitivity, and control. Here are the three dominant delivery models:

Elastic compute via managed services

Hyperscalers (Azure, AWS, GCP) offer on-demand GPU access through APIs, SDKs, and managed runtimes. You request resources, and the platform allocates GPU capacity behind the scenes whether from Nvidia or proprietary chips like AWS Trainium or Microsoft Maia.

For PMs, this model supports rapid iteration, dynamic scaling, and global reach—ideal for GenAI workloads with variable demand (e.g., consumer chatbots, seasonal agent tasks). The trade-off? Less control over placement, latency, and architecture. You're building on their terms.

Bare-metal GPU access (with GenAI in mind)

Providers like CoreWeave offer bare-metal access, meaning you lease the physical GPU server directly, without virtualization layers. This gives you full control over memory allocation, process scheduling, and fine-tuning environments. For model builders, training scenarios and inference-heavy agentic systems, this model delivers lower latency, higher throughput, and cost-efficiency at scale.

But it requires more operational maturity—provisioning, deployment scripts, and usage forecasting fall on you. CoreWeave's edge comes from rapid provisioning—often deploying H100s and GB200s in hours, compared to the weeks it might take on hyperscalers. Bare metal delivers control but exposes you to cost volatility and scheduling risk if workloads aren't predictable.

Integrated software toolchains

Chipmakers (e.g., Nvidia) bundle hardware with software: CUDA for GPU programming, Triton for optimized inference, and NIMs (Inference Microservices) for production-ready model containers. These speed up deployment but increase lock-in. What simplifies scaling today might limit portability tomorrow.

How value is captured

Infrastructure providers don't monetize intelligence, they monetize access to the engine behind it. Their business models are built around consumption, predictability, and platform stickiness—all of which directly shape your product's cost structure.

Usage-based billing (the core meter)

Most infra providers charge by GPU hours, token volume, or model runtime. Every model call—whether a quick summarization or a multi-step agent task—adds to your bill. As usage scales, infrastructure becomes your second P&L. Margins can erode quickly if pricing doesn't track closely to perceived value.

Reserved capacity & enterprise commitments (the forecast gamble)

To lower per-unit costs, providers offer reserved compute (e.g., provisioned throughput units, or PTUs, from Microsoft) or multi-year bundles (e.g., Microsoft Azure Consumption Commitment, or MACC, agreements). These provide pricing stability, but at a cost: if you don't use what you commit to, it becomes a sunk cost. For PMs, this means your demand forecast isn't just a spreadsheet, it's a business model risk.

Bundled ecosystem revenue (the full stack play)

Infra vendors increasingly capture value across the stack. Nvidia doesn't just sell GPUs, it monetizes SDKs and inference accelerators. Cloud platforms bundle compute spend with monitoring, orchestration, and security. The more integrated you become, the more billing surfaces they can tap.

Platform lock-in (the long game)

Long-term value capture often comes from ecosystem stickiness. When your backend depends on platform-specific runtimes, SDKs, or provisioning models, switching becomes non-trivial. That gives providers pricing leverage, especially when bundled with performance SLAs or support tiers. Lock-in isn't inherently bad—but it must be a *strategic choice*, not an accidental outcome.

Infrastructure isn't just an enabler; it shapes the economics of every GenAI interaction. Your choice of provider isn't just about performance. It's a long-term bet on cost structure, scalability, and control.

Case study: CoreWeave – Infrastructure as a competitive edge

For years, GenAI infrastructure was dominated by hyperscalers like Azure, AWS, and Google Cloud. But in the post-2023 wave, CoreWeave emerged as a GPU-native cloud built specifically for GenAI.

Originally a crypto mining startup, CoreWeave pivoted hard into GenAI, differentiating not on breadth, but on depth: fast provisioning, bare-metal performance, and developer-first APIs. While hyperscalers throttled quotas and priced conservatively, CoreWeave became the go-to for model builders and agentic startups who needed elasticity without compromise.

By April 2025, it was powering workloads for OpenAI, Mistral, Inflection AI, and Runway. The real breakthrough came in March 2025, when OpenAI struck an $11.9B deal to name CoreWeave a preferred infrastructure partner, backed by a $350M investment.

This move signaled the rise of a new category in GenAI infrastructure, specialized GPU-native providers distinct from general-purpose hyperscalers like Microsoft, AWS, Google, and Oracle.

Its strategic edge? Tight alignment with Nvidia, including early access to Nvidia's latest superchips and joint investments. This hardware-software synergy helped CoreWeave deliver next-gen GPUs to customers faster than the competition.

Here is what GenAI PMs can take from this example:

- Specialized infra isn't just cheaper, it can unlock new product tiers. If your product hinges on real-time inference or fine-tuning, infra tuned for GenAI—not general-purpose cloud—may be a game changer.
- Bare-metal isn't just a technical choice, it's a pricing and control lever. It can improve margins at scale but adds operational complexity you must plan for.
- You don't need to build chips to shape economics. Smart packaging, provisioning, and delivery—done right—can be just as powerful.

In the next archetype, we'll explore how foundation model providers create value, compete, and monetize by delivering intelligence as a service.

Archetype 2: Foundation Model Providers

Where Intelligence Is Created and Moats Are Harder to Defend

If GenAI infrastructure is the engine, foundation models are the fuel—the cognitive layer behind generation, reasoning, summarization, and decision-making. But as we discussed earlier, model performance alone no longer ensures defensibility.

Three forces are converging:

- Commoditization of core capabilities
- Model interchangeability through abstraction layers like LangChain and Semantic Kernel
- Downward pressure on API pricing—driven by intense competition, enterprise cost sensitivity, and the rise of open-weight alternatives

The result? Model providers create immense value but increasingly struggle to defend it. They're under pressure to differentiate not just on quality, but on business models, control, integration depth, and ecosystem reach. In this archetype, we'll examine how leading providers—OpenAI, Anthropic, Meta, Cohere, Mistral—are adapting to this shift, and what PMs need to weigh when deciding who to build on.

The result? Model providers still create immense value but struggle to protect it. Differentiation now hinges on more than just output quality. It's about business model control, integration depth, and ecosystem strength.

How is value created

Foundation model providers create value by transforming vast datasets and computational resources into general-purpose intelligence delivered as code, predictions, completions, or multimodal outputs.

But in today's market, value isn't just about training a capable model—it's about enabling it to support real-world use cases, safely and efficiently. They do this through the following methods:

- Pretrained intelligence: Large models trained on broad datasets to enable a wide range of reasoning, writing, and coding tasks.
- Specialized tuning: Instruction alignment and fine-tuning for vertical use cases (e.g., legal drafting, medical summarization).
- Multimodal capability: Support for vision, text, speech, and structured input/output.
- Inference infrastructure: Hosting via APIs, private endpoints, or marketplaces to make intelligence accessible at scale.

The model is just the starting point. The real value lies in making it usable, trustworthy, and adaptable.

How is value delivered

Model providers deliver value through accessible, scalable, and flexible interfaces that let developers and enterprises integrate intelligence into workflows. They offer three core delivery paths:

APIs and SDKs (speed to market)

Most teams consume model intelligence via cloud APIs—hosted endpoints with tokenized access, like OpenAI's GPT or Claude from Anthropic. These abstract away infra complexity and allow fast integration but come with usage-based pricing and limited control.

Managed fine-tuning services (task alignment)

Providers like OpenAI, Mistral, and Cohere allow you to customize a base model via hosted fine-tuning platforms—often bundled with evaluation tools, data guards, and SLAs. This supports higher accuracy on domain-specific tasks but adds cost and operational dependency.

Private/VPC deployments and licensing (enterprise control)

For regulated or high-security use cases, models are deployed inside private environments—either through self-hosted weights (e.g., Llama, Mixtral) or VPC containers. These setups offer maximum control, performance tuning, and data privacy but require significant infra readiness and legal overhead.

At this layer, how intelligence is packaged—not just how it performs—shapes the developer experience and enterprise adoption curve.

How is value captured

Foundation model providers monetize intelligence through a range of mechanisms, each balancing control, scalability, and integration. Here are the dominant strategies:

Token-Based APIs (pay as you go):

The default for most hosted models. Providers like OpenAI and Anthropic charge per token generated or processed, often with tiered pricing by model family (e.g., GPT-4 Turbo vs. GPT-4). This aligns revenue with usage but can create cost friction for products with high interaction volume.

Managed fine-tuning (customization as a premium):

Providers monetize premium support for task-specific tuning. These offerings often include SLAs, evaluation tooling, and hosting. It's less about the model weights and more about the service layer that helps enterprises get predictable performance.

Model licensing (control at a cost):

Open-weight providers like Meta or Mistral allow companies to self-host model-charging fees based on user thresholds (e.g., Meta's 700M MAU cap) or redistribution rights. This gives enterprises flexibility but introduces legal and operational overhead.

Private/VPC deployments (security as a service):

Some providers offer deployment inside a customer's cloud or VPC—serving industries where data sovereignty, latency, and auditability are non-negotiable. The premium here is trust.

Marketplace listings (distribution with a cut):

Models listed on platforms like Azure AI Foundry, AWS Bedrock, or Hugging Face capture value via metered usage or revenue share. These channels increase discoverability but often compress margins and limit customization.

Strategic insights for GenAI PMs

Whether you're shaping a model offering or building on top of one, your job as a GenAI PM is to turn provider capabilities into product advantage. That starts with looking beyond raw performance. Focus on how the model enables differentiated UX like chatbots that feel human or agents that anticipate user needs. Explore co-development opportunities with providers to integrate custom workflows, build trust, and increase adoption.

In regulated industries, compliance can't be an afterthought. Choose providers with robust safety features and build guardrails into your product. A single misstep can be costly. If you're distributing via marketplaces, invest in branding. Limited customization can blur product identity, so positioning becomes everything.

Finally, don't dismiss open-weight models. They offer cost flexibility but require investment in optimization, evaluation, and deployment maturity. Turning "free" into a competitive advantage takes work.

Foundation model providers power the cognitive core of GenAI but in a commoditizing market, the right provider isn't just the most accurate. It's the one aligned with your pricing model, compliance risks, and go-to-market strategy. Choosing a model means choosing your trade-offs.

Case study: Anthropic and the business of trust

When Anthropic launched Claude, it didn't chase model benchmarks or open-source hype. Instead, it bet on something harder to define—but just as valuable: *trust.*

Founded by former OpenAI researchers, Anthropic bet that in high-stakes environments—healthcare, legal, finance—predictability and safety would be differentiators. Claude was designed to behave reliably, safely, and interpretably in sensitive workflows.

Rather than compete on token price or weight-sharing, Anthropic built a trust-first API business, monetizing across multiple layers:

- A developer-friendly API with fast latency, high context windows, and flexible usage tiers
- Claude Pro, a premium subscription tier targeted at small and medium-size enterprises (SMEs) and prosumers
- Deep integrations with Slack, Notion, and Zoom—powering real-time summarization, task gen, and structured output
- A growing library of hosted agents and plug-ins designed for retrieval-heavy, compliance-aware workflows

By 2025, Claude had become the preferred model for institutions where auditable, explainable, and repeatable output mattered more than raw capability or cost.

Here is what GenAI PMs can take from this example:

- Trust is monetizable. Claude proves that safety and predictability can drive enterprise willingness to pay.

- UX + vertical fit = pricing power. Claude excels in document-centric workflows (e.g., contract review, claims processing) where clarity and structure matter more than creativity.
- APIs are just the wedge. Anthropic's real moat is ecosystem depth. PMs should think similarly—how your agents, integrations, and support layers reinforce each other over time.

In the next archetype, we'll explore how platforms and marketplaces turn model intelligence into products—shaping how GenAI gets distributed, discovered, and adopted at scale.

Archetype 3: Platforms & Marketplaces

Where Intelligence Meets Distribution and Productization Begins

If foundation models are the fuel, platforms and marketplaces are the delivery network—connecting model providers to developers, enterprises, and real-world use cases. Sitting between raw intelligence and end-user experiences, this layer defines how AI is discovered, customized, orchestrated, and scaled.

This archetype includes orchestration frameworks, inference-serving platforms, and model marketplaces like Azure AI Foundry, Hugging Face, AWS Bedrock, and others. These platforms simplify access, speed up integration, and offer plug-and-play intelligence, reshaping how GenAI is consumed and monetized.

For product managers, understanding this layer is critical. It shapes your speed to market, paths to customization, pricing levers, and how much control—or dependence—you'll retain.

How value is created

Platforms and marketplaces create value by removing friction from the GenAI delivery chain. They turn raw model intelligence into usable components, often bundled with observability, fine-tuning support, security, and orchestration frameworks. Their core value lies in the following:

- Aggregation: Curated catalogs of proprietary, open-weight, and fine-tuned models across modalities—text, vision, speech, and more.
- Abstraction: Simplified APIs and orchestration layers that mask complexity—model routing, tool use, and agent execution.

- Acceleration: Built-in tools for fine-tuning, evaluation, benchmarking, and deployment—cutting model onboarding from weeks to minutes.
- Standardization: Quality filters, compliance guardrails, and ecosystem alignment (e.g., support for LangChain, Semantic Kernel, ONNX).

By wrapping intelligence with usability and governance, platforms turn potential into product-ready capability.

How value is delivered

Delivery happens through four strategic mechanisms:

- Unified APIs: Platforms like Bedrock and Foundry offer a single interface for accessing multiple models. Developers can swap Claude for Mixtral or Llama with minimal code changes, thanks to standard interfaces and orchestration patterns.
- Pre-built templates and fine-tuned workflows: Azure AI Foundry, for example, provides no-code/low-code flows for fine-tuning and deploying models into enterprise apps. This removes bottlenecks for PMs who need to validate ideas without standing up infra.
- Hosted agent infrastructure: Many platforms now support lightweight agentic patterns such as tool calling, memory, and multi-model orchestration. While still evolving, this infrastructure lets teams prototype agent behavior before committing to full autonomy. (We'll explore the full business implications of agentic AI in a later section.)
- Marketplace distribution: These platforms also serve as distribution channels. Fine-tuned models, vertical agents, and custom endpoints can be published, monetized, and shared—accelerating adoption and enabling long-tail innovation.

How value is captured

Unlike model providers, platforms don't always monetize the intelligence directly—they monetize the packaging, tooling, and distribution around it. Here are the dominant monetization levers in play:

Usage-based APIs:

Most platforms charge per inference, either token-based (model usage) or request-based (agent/task invocation). These fees scale with user activity and are the primary revenue stream for public endpoints.

Marketplace commissions:

When third-party models or apps are listed, platforms typically take a 10–30% commission per transaction—depending on the pricing tier and partner agreement.

Hosted infrastructure & platform tiers:

Many platforms offer hosted inferencing, memory, or orchestration features (like Semantic Kernel or vector search) as part of paid service tiers. This includes support for fine-tuning, sandboxing, agent persistence, or retrieval pipelines.

Enterprise reserved capacity (e.g., PTUs):

Just like infrastructure providers, platforms increasingly offer reserved compute options—such as provisioned throughput units (PTUs)—that guarantee inference capacity for enterprise deployments. This supports SLAs and predictable pricing for production workloads.

Ecosystem lock-in (strategic, not always monetized):

Some platforms create long-term leverage through SDKs, orchestration frameworks, or proprietary runtimes that embed deeply into your stack. This isn't always monetized directly but it increases switching costs, giving the platform pricing power over time.

Strategic insight for GenAI PMs

As a GenAI PM, whether you're shaping a platform's capabilities or building your product on top of one, your success hinges on how well you leverage the platform's strengths—while designing around its limitations.

Experiment across models, fast. Choose platforms that let you test multiple models quickly through unified APIs. This flexibility enables you to experiment across providers—swapping in a more accurate or cost-effective model—without stalling your product's release cycle.

Prioritize observability in tuning. When fine-tuning is part of your strategy, favor platforms with strong evaluation and monitoring features. Customization is only valuable if you can measure its long-term performance and ensure it adapts as user needs evolve.

Match infra choices to compliance needs. If you're targeting regulated markets, opt for platforms that offer enterprise-grade compliance support and reserved capacity. This ensures you can scale reliably while meeting strict audit, residency, and latency requirements.

Stand out in the marketplace. Marketplace distribution can expand your reach, but visibility is limited. A differentiated user experience like seamless agentic workflows that solve high-friction problems, can set your product apart in a crowded catalog.

Future-proof with modular integrations. Lock-in isn't always bad, but it should be intentional. Using standard orchestration frameworks like LangChain or Semantic Kernel gives you the flexibility to evolve your architecture if better options emerge.

Platforms and marketplaces are the connective tissue of the GenAI ecosystem. They don't build the models but they determine how quickly, safely, and widely those models reach the world. For product managers, choosing the right platform means balancing *speed to market, customization pathways, pricing structures,* and *long-term control over your AI stack.*

Case study: Hugging Face—platform power without owning the model

Hugging Face started as a hub for open-source NLP models and evolved into what many now call the *"GitHub of AI."* But it didn't stop at model sharing, it built a business by turning aggregation, deployment, and collaboration into monetizable services.

Instead of chasing frontier model leadership, Hugging Face positioned itself as the connective tissue of the open-weight ecosystem. Its offerings—like inference endpoints, private model hubs, and fine-tuning flows—are wrapped in developer-friendly APIs and backed by integrations with AWS, Microsoft, and Meta.

By 2025, enterprise products like Inference Endpoints (for secure deployment) and AutoTrain (for low-code fine-tuning) blurred the line between developer enablement and infrastructure-as-a-service.

Its monetization strategy reflects the archetype's diversity:

- Token-based pricing for API-driven inference

- Enterprise hosting fees for VPC deployments and private hubs
- Fine-tuned workflows offered as managed services (e.g., AutoTrain)
- Marketplace-style revenue share from third-party model listings
- Cloud distribution deals embedding Hugging Face tooling into partner ecosystems

Here is what GenAI PMs can take from this example:

- *Middleware is a business model.* Hugging Face didn't build the models or own the GPUs—it monetized enablement, making it easier for others to build and scale.
- *Curation builds trust.* Features like model cards, benchmark scores, and responsible AI filters differentiate it in a crowded marketplace.
- *Customization doesn't have to be complex.* Tools like AutoTrain make tuning accessible—transforming enterprise control into a repeatable, service-led motion.
- *Ecosystem-first thinking drives adoption.* By prioritizing integrations (LangChain, PyTorch, Azure, AWS), Hugging Face became more than a platform—it became a movement.

While Hugging Face scaled through broad access and openness, the next wave of platforms is going deep—designing GenAI for specific domains and regulations. These vertical builders aren't just using AI—they're redefining it for healthcare, law, finance, and manufacturing.

Next up: Vertical AI Builders—where domain expertise meets customization to create high-value, defensible solutions.

Archetype 4: Vertical AI Builders

Where General Intelligence Meets Domain Expertise

In a landscape dominated by general-purpose models and open APIs, one question keeps surfacing: *Can GenAI do my job?* For many industries—like healthcare, finance, legal, and manufacturing, the answer depends not just on intelligence, but on context. That's where *vertical AI builders* come in.

These companies don't just use models, they tailor them. By combining domain expertise, workflow integration, and industry compliance, they build solutions that speak the language of a doctor, a lawyer, or a compliance officer.

Instead of racing to build the most powerful model, they focus on *fit*—designing systems that generate real outcomes: faster diagnoses, more accurate contract reviews, better audit readiness. Their value isn't just in what the model *outputs*, but in what the product *achieves*.

For GenAI PMs, this archetype is a shift in mindset. It's not just about tokens and models, it's about *jobs to be done, workflow impact,* and *customer-specific* ROI. Let's explore how these players create, deliver, and capture value.

How value is created

Vertical AI builders create value by aligning GenAI with industry-specific knowledge, regulations, and workflows—translating raw AI capability into practical business outcomes. Rather than chasing general performance, they optimize for domain relevance and real-world usability.

They do this through:

- **Domain-specific fine-tuning:** Using curated datasets to adapt foundation models to sector-specific language and formats like legal citations, patient notes, or tax forms
- **Workflow integration**: Embedding GenAI directly into professional tools and enterprise platforms whether it's a radiologist's workstation or a financial CRM.
- **Compliance alignment**: Designing systems that respect HIPAA, GDPR, FINRA (the Financial Industry Regulatory Authority), or other regulatory frameworks, often with built-in audit trails and explainability features.

In this archetype, value shifts from *what the model can do* to *how well the system fits*— where clarity, traceability, and regulatory alignment are just as important as raw accuracy.

How value is delivered

Value is delivered through deeply integrated products designed for professional grade use cases, not just as APIs or chatbots, but as seamless extensions of industry workflows. These builders deliver GenAI through the following:

- **AI-powered SaaS tools**: Products tailored for legal research, contract review, clinical documentation, tax prep, or underwriting built around specific jobs to be done.

- **White-labeled copilots**: Embedded AI assistants that plug into EHRs, accounting platforms, or industry-specific CRM tools.
- **End-to-end vertical platforms**: Full-stack solutions with data ingestion, reasoning, human-in-the-loop review, and compliance-ready reporting

Often, these builders *don't expose the model at all*. What users see is a product that understands their job and just gets it done.

How value is captured

Vertical AI builders monetize by tying revenue to *outcomes, task completion,* and *workflow depth*—often blending multiple strategies to align with customer value. Here's how value is captured:

- **Vertical SaaS licensing**: Subscription models priced per user or per seat, similar to traditional enterprise software but with GenAI at the core.
- **Per-task or outcome-based pricing**: Charging per contract reviewed, diagnosis suggested, or claim triaged, directly linked to measurable ROI.
- **API gateways for industry data**: Monetizing structured, enriched datasets—such as annotated legal cases or de-identified medical records—via API or bundled with proprietary models.
- **Professional services & customization**: For complex enterprises, revenue comes from tailoring models to local jurisdictions, languages, or regulatory frameworks.

Many also partner with cloud providers for hosting and monetization while retaining control over workflows, UX, and pricing—ensuring consistent delivery and regulatory alignment.

Strategic insight for PMs

As a GenAI PM working with or building a vertical AI solution, this archetype shifts your focus: *from raw model performance to workflow fit, trust, and domain depth*.

Start with the user's job, not the model's output. Fine-tuning alone won't make your product useful, design for how the work gets done. If a doctor needs fast, compliant diagnostics, your system should meet that need naturally, without retraining behavior.

Drive adoption through usability. Build interfaces that feel native to your users' existing tools—and offer intuitive onboarding, especially in technical or regulated fields.

Make compliance a feature, not a burden. Auditability, traceability, and transparency can be differentiators in industries where trust is everything.

Rethink distribution. In vertical markets, traction often comes from word-of-mouth, partnerships, or niche marketplaces not mass adoption channels. Finally, *Treat data as your moat.* Curated, domain-specific datasets and feedback loops compound over time, improving accuracy and reinforcing product stickiness.

In vertical markets, *intelligence without context is noise.* What vertical AI builders create is signal—solutions that are not just smart, but relevant. The next wave of GenAI value won't come from bigger models but from making AI deeply useful in real work, for real people.

Case study: Harvey—vertical AI for the legal profession

In the world of GenAI, general-purpose assistants might write code or draft emails. But in law—where trust, precision, and confidentiality are non-negotiable—Harvey stood out by going deep, not broad.

Launched in 2022, Harvey set out to build an *AI-native platform for lawyers*, not just an AI that could process legal text. It combined foundation models with legal-specific fine-tuning, retrieval pipelines, and secure deployment options to help lawyers draft memos, analyze contracts, respond to RFPs, and conduct legal research.

Its impact was immediate. Harvey signed major firms like Allen & Overy and PwC Legal, embedding itself as a trusted copilot for elite legal teams. What made it defensible wasn't just performance, it was precision, privacy, and integration.

Key differentiators included:

- Fine-tuned LLMs trained on proprietary legal corpora—statutes, case law, firm-specific precedents
- Secure deployments via VPCs or air-gapped cloud environments
- Industry certifications supporting confidentiality, auditability, and compliance
- Tight integration with legal tools like contract negotiation platforms and research databases

Harvey's *monetization model mirrored its customers' needs.* It offered per-seat licensing for law firms, with pricing tiers based on practice areas and data volume. Larger clients received enterprise-grade packages that included compliance guarantees, SLAs, and dedicated support. For firms requiring air-gapped or on-prem solutions, Harvey delivered custom deployments tailored to stringent security and residency requirements. In newer pilots, it experimented with outcome-based pricing—charging based on time saved or the volume of research automated—linking revenue directly to measurable productivity gains.

Here's what GenAI PMs can learn from this example:

- *Customization is the product.* Harvey didn't try to outcompete OpenAI on model architecture—it out-packaged it with legal-grade tuning and user experience. In vertical AI, context beats scale.
- *Data is the moat.* Harvey's advantage wasn't in novelty, but in access. Its legal performance came from proprietary data—court filings, firm documents, annotated case summaries—that general models can't touch.
- *Monetization follows trust.* Law firms paid more for reliability, privacy, and explainability. Harvey earned that trust by designing for regulation, not just raw accuracy.
- *Go-to-market is vertical-first.* Harvey wasn't trying to serve every profession. It served one customer deeply—the lawyer. By focusing on their workflows, tools, and pain points, it built credibility, adoption, and long-term defensibility.

Archetype 5: AI-Native Applications & Ecosystem Aggregators

Where GenAI Reaches Users

If infrastructure powers GenAI and models provide intelligence, AI-native applications are where it becomes visible—*and valuable*. These products aren't just GenAI-enabled; they're GenAI-native, built from the ground up to turn model capability into user-facing functionality.

From copilots and assistants to AI-powered search, writing, and decision tools, these apps deliver real utility—one task at a time. Some go deep within a vertical (e.g., GitHub Copilot for developers). Others aim for horizontal reach (e.g., Perplexity for research, Notion AI for productivity). A few even evolve into *ecosystem aggregators*—pulling in models, tools, memory, and user context across domains (e.g., ChatGPT with GPTs and the Assistants API). Regardless of scope, these apps sit closest to the user and that proximity is power.

How value is created

AI-native applications create value by *translating intelligence into workflow*. They don't just expose a model, they productize cognition: summarizing content, generating ideas, answering questions, guiding actions. The innovation isn't just technical, it's experiential.

This value is created through the following:

- **Multimodal orchestration:** Blending text, code, image, and speech across input/output layers
- **Retrieval and context injection:** Grounding outputs using user data or external knowledge via RAG pipelines and APIs
- **Personalization engines:** Adapting responses and UX through behavior signals and contextual memory
- **Cross-model composition:** Routing requests across models (e.g., Claude, GPT-4, Mixtral) based on cost, latency, or task fit

These companies don't win by owning the full stack, they win by *owning the experience*.

How value is delivered

AI-native applications deliver value through *high-utility user experiences* where GenAI is often invisible but indispensable. These products abstract away prompts, models, and tokens, surfacing only what users care about: fast, relevant, and trustworthy outcomes with minimal friction.

Delivery takes many forms. *Conversational UIs* like ChatGPT, Grok, and Perplexity center around chat-first interactions, often enhanced by memory, plug-ins, and light agent workflows. *Embedded experiences*—such as Notion AI or Microsoft Copilot—bring GenAI into existing productivity tools, enabling seamless task completion within familiar interfaces. *Developer-facing tools* like GitHub Copilot integrate directly into IDEs, accelerating common workflows with minimal disruption.

The best applications collapse time-to-value, offering clarity, control, and output with minimal input. Increasingly, this includes *agentic behaviors*—where applications don't just assist, but act on users' behalf. Early signs of autonomy—like inbox triage, multi-step task execution, or intelligent retrieval—are already emerging, hinting at what's next: a shift from passive tools to proactive agents. *(We'll explore that shift—and its business model implications—in the next section on Agentic AI.)*

How value is captured

AI-native applications and ecosystem aggregators capture value through a blend of consumer-friendly pricing and infrastructure-aware economics.

Many start with a *freemium model* offering limited features or queries for free, then monetizing through premium subscriptions. Products like ChatGPT Plus and Notion AI Pro follow this path, converting usage into recurring revenue once users experience value.

Others adopt *usage-based pricing*, charging by the number of API calls, document generations, or user seats especially in enterprise or team contexts. Some embed GenAI capabilities directly into *bundled SaaS offerings*, as seen with Microsoft 365 Copilot or Canva's Magic Write, making AI a natural part of existing subscriptions.

More advanced applications explore *in-app monetization*, enabling users to purchase plug-ins, custom agents, or GPTs through marketplace models, turning the app itself into a revenue platform.

The closer you are to user outcomes, *the more pricing power you command*. But every interaction—every token, every task—adds to your cost base. *Inference economics matter just as much here as they do for model and infra providers*. Sustainable monetization requires balancing delight and discipline, ensuring value delivered exceeds cost incurred.

Strategic insights for PMs

As a GenAI PM—whether you're building an AI-native app or integrating one into a broader ecosystem—this archetype highlights one core truth: *user proximity drives product success*.

Prioritize experience over architecture. In a world where many models perform similarly, *UX is your product*. Seamless, intuitive interactions are what differentiate and drive retention.

Distribution matters more than invention. The best GenAI apps don't always use the most advanced models, but they scale through *growth loops, integrations, and ecosystem effects*. Focus on building the channels that amplify reach.

Inference is your COGS. Every AI output has a cost. Sustainable business models depend on how well you manage model usage, routing, and margins. Optimize for value delivered per token spent.

If your product supports agentic workflows, build for observability. Users need clarity when tasks span multiple steps or tools. Visibility, fallback paths, and trust safeguards are critical to maintaining confidence.

Don't just build an app, build an ecosystem. The most defensible GenAI apps evolve into platforms, supporting plug-ins, extensions, or embedded agents that drive ongoing usage and developer-led growth.

Case study: ChatGPT—from model showcase to AI app platform

When OpenAI launched ChatGPT in late 2022, it began as a simple interface to showcase GPT's capabilities. But it quickly became something more: a product that hit 100 million users in two months and went on to become OpenAI's fastest-growing revenue driver.

By 2025, ChatGPT had evolved into a full-fledged AI-native platform—powering not just conversations, but apps, workflows, and a growing ecosystem. While the free tier remained widely used, monetization emerged through multiple channels. *ChatGPT Plus* introduced a $20/month subscription offering access to GPT-4 Turbo, faster responses, and priority features. The *GPTs Marketplace* enabled creators to build and monetize specialized GPTs for everything from legal summaries to fitness plans.

OpenAI also launched an *Assistants API* for developers to embed GPT-powered workflows into external tools. And through *enterprise licensing*, companies began integrating ChatGPT into customer support, internal copilots, and vertical SaaS products.

What's unique is how OpenAI transformed a model interface into an *ecosystem*—positioning ChatGPT not just as a standalone product, but as a distribution layer for intelligence.

Here's what GenAI PMs can learn from this example:

- *Great models need great experiences to shine.* ChatGPT's success wasn't just about GPT-4—it was about packaging intelligence with intuitive UX, feedback loops, and reliability.
- *User trust is a product moat.* Features like memory, history, and privacy controls turned usage into habit and habit into loyalty.
- *Monetization can grow from the edge inward.* Revenue started with individuals, expanded into enterprises, and now supports a creator economy through GPTs and plug-ins.
- *AI-native UX drives adoption.* ChatGPT became the default GenAI interface for millions by making intelligence feel conversational, responsive, and personal.

ChatGPT shows how AI-native apps and ecosystem aggregators bring GenAI directly to users—transforming intelligence into daily workflows through seamless design and extensible platforms. But as these apps increasingly adopt agentic capabilities—like OpenAI's *Operator* automating browser tasks—they're beginning to unlock a new phase of product innovation: one defined by autonomy, adaptability, and intelligent delegation.

Let's now explore how *agentic AI* is transforming how businesses create, deliver, and capture value and what that means for the next generation of GenAI products.

Agentic AI: Transforming Business Models

Imagine an AI that doesn't just respond to prompts but behaves more like a capable colleague—able to reason, plan, and carry out complex tasks without constant supervision. That's the promise of *agentic AI*. And as of mid-2025, it's no longer theoretical. Businesses are already embedding agents into key operations, fundamentally reshaping how value is created, delivered, and captured.

In this section, we'll explore how agentic systems are driving automation, enabling personalization at scale, supporting strategic decision-making, and opening new paths for monetization.

We'll also look ahead at how this shift could redefine not just productivity, but the structure of business itself.

How value is created

Rewriting how work gets done

Agentic AI enhances how businesses generate value not by replacing human input, but by enabling new forms of output altogether. From logistics and HR to healthcare and product development, agents are stepping into roles once reserved for specialists or operators.

Automation with precision, not just speed

In logistics, agentic systems are already taking control of complex workflows. Microsoft's supply chain agents, for instance, don't just assist—they actively manage delivery routes, inventory levels, and real-time coordination across vendors. What used to require a room full of analysts now runs continuously in the background.

Meanwhile, Workday's Illuminate platform automates HR workflows such as onboarding, leave approvals, and benefits updates—executing them with a pace and consistency that would overwhelm human teams. These aren't marginal improvements, they reflect a structural shift in how operational work gets orchestrated through AI autonomy.

Personalization grounded in real-world context

In healthcare, personalization isn't just a UX goal, it's often a clinical requirement. Tempus, for example, uses agentic systems that analyze patient history, genomic data, and clinical guidelines to recommend individualized treatment paths. These AI-driven suggestions enhance both speed and relevance, making it easier for clinicians to deliver targeted care at scale. The system doesn't replace doctors, it augments their ability to act quickly and confidently in complex cases.

The same principle applies outside healthcare. In finance, retail, and education, agents are learning from customer data to deliver product recommendations, proactive service alerts, or contextual guidance that feels like it was handcrafted—without actually being so. The goal isn't hyper-personalization for its own sake; it's relevance, delivered when and where it matters.

Human–AI collaboration with clearer boundaries

Agentic systems aren't replacing humans; they're reshaping how we work with machines. As we covered in the book earlier, GitHub Copilot is a familiar example: it suggests code snippets, tests, and documentation in real time, allowing developers to stay focused on architectural thinking rather than syntax.

In creative and strategic roles, agents are increasingly taking over repetitive or compliance-driven tasks—freeing up teams to focus on high-leverage decisions. The outcome isn't just faster execution, it's different thinking, because human time is no longer consumed by busy work.

Learning loops that actually loop

What sets agentic systems apart from traditional automation is their capacity to improve over time. Whether it's a routing agent optimizing delivery paths or a support agent refining its escalation logic, these systems aren't fixed assets— they're dynamic ones.

Through structured feedback, reinforcement learning, and telemetry, agents can refine their performance with each task they complete. In practice, this means the cost of experimentation falls, and the upside of iteration increases—a favorable curve for any product team aiming to compound value over time.

Agentic AI isn't about handing the keys to a machine. It's about redesigning workflows to get better results—consistently, safely, and at scale. Businesses that recognize this shift aren't just automating; they're architecting new ways to operate, compete, and grow.

How value is delivered

From static tools to dynamic, personalized services

Once value is created, the next challenge is delivering it, getting the right outcome to the right user at the right time. Agentic systems change this equation. Instead of relying on static interfaces or rigid workflows, they enable services that adapt in real time and act on user intent. For GenAI product managers, this means designing experiences that feel seamless and responsive without demanding constant input.

Salesforce's Agentforce, for instance, monitors backend signals like delivery delays or churn risk—and proactively intervenes before users even notice a problem. ChatGPT's Operator automates browser-based tasks such as booking appointments or submitting forms, turning intent into action with minimal friction.

Expedia's AI Concierge (Romie) adjusts travel plans in real time, helping users rebook flights when weather causes delays. In IT, Power Design's HelpBot flags maintenance issues before they impact operations, scaling support while improving reliability.

At the orchestration layer, Microsoft Copilot Studio connects agents across systems and tools, enabling cross-team workflows. Azure AI Foundry complements this with infrastructure for memory, tool chaining, and secure hosting—letting agents persist context and manage multi-step tasks effectively.

This shift isn't just about speed—it's about narrowing the gap between goal and outcome. These systems don't just assist; they participate. And as agentic delivery becomes more common, it's redefining what users expect from digital products.

How value is captured

From subscriptions to outcomes

AI-native apps have traditionally relied on familiar SaaS playbooks—subscriptions, usage-based pricing, or bundled tiers. But as these products grow more autonomous, value capture is starting to evolve.

Some enterprise service providers are experimenting with outcome-based models. Deloitte and EY, for instance, are piloting fee structures tied to actual impact— cost savings, efficiency gains, or regulatory compliance delivered by AI agents. In SaaS, startups are testing pricing tied to specific agent outcomes: meetings booked, support cases resolved, or tasks completed without human input.

Even in regulated sectors, this shift is gaining traction. Anthropic's secure agents, though currently priced by usage, hint at a future where risk flagged or cost avoided could drive monetization. Salesforce's Agentforce could move toward per-resolution pricing. ChatGPT Pro's flat monthly fee may shift one day to pricing based on tasks completed or time saved.

For GenAI PMs, this evolution introduces new responsibilities:

- Design for outcomes. Ensure your agents consistently deliver measurable, valuable results.
- Track and improve. Build observability and feedback loops that tie success to specific metrics.
- Experiment with pricing models. Explore ways to align cost with value delivered whether that's saved time, higher output, or fewer human escalations.
- Mitigate risk in sensitive domains. Be ready to prove reliability, fairness, and compliance especially in healthcare, finance, and education.

You're not just monetizing features, you're monetizing outcomes. That changes how you build, measure, and price what you ship.

Looking Ahead

From agentic tasks to ecosystem strategy

Agentic AI is no longer just a productivity layer, it's becoming a strategic force. As this shift accelerates, GenAI product managers won't just build features; they'll shape how businesses evolve, collaborate, and grow.

Tomorrow's agents won't just manage tasks, they'll help orchestrate entire ecosystems. Imagine agents that track global market signals and proactively recommend product pivots. Or that partners with third-party agents across companies to co-develop solutions—coordinating logistics, reallocating resources, or identifying shared opportunities, all without human initiation.

This isn't science fiction. Early signs are already emerging. Platforms like Credo AI, which focus on policy enforcement and AI governance, hint at a future where agents don't just work, they uphold standards, negotiate trade-offs, and enforce transparency across domains like finance, healthcare, and education.

For business models, that evolution will look like this:

- *How value is created:* Moves from automating workflows to co-creating strategies. Agents help uncover new market opportunities or optimize operations at the ecosystem level, not just within a single company.
- *How value is delivered:* Becomes more distributed and collaborative. Agents interact across organizational boundaries, delivering outcomes through shared systems and networks, not siloed tools.
- *How value is captured:* Shifts toward shared outcomes. Businesses and agents may earn value based on contribution to broader goals like efficiency gains across a supply chain, or measurable impact on societal challenges.

To lead in this future, you'll need more than technical fluency. You'll need to think like a systems architect—designing for interoperability, ethical alignment, and cross-agent collaboration. Because in the next decade, your greatest innovations may not come from what your agent does alone but from what it enables others to do with it.

Conclusion

In this chapter, we've explored five archetypes that define how GenAI creates, delivers, and captures value in the real world—from infrastructure providers and model builders to orchestration platforms, AI-native applications, and agentic systems. Each archetype revealed a different lever in the AI value chain and a different way product managers can shape the future.

We saw how *infrastructure providers* turn raw compute into scalable services, anchoring every GenAI product in performance and cost. *Model builders* showed us how advances in pretraining, fine-tuning, and evaluation unlock new capabilities—and new tradeoffs. *AI frameworks and orchestrators* introduced the logic layer, where reasoning, memory, and tool use are fused into workflows. *AI-native apps and ecosystem*s brought these ingredients together into user-facing experiences, turning intelligence into daily utility. And finally, *agentic systems* pointed to what's next: a future where AI doesn't just assist—it acts, learns, and collaborates.

If there's one thread across all five, it's this: value in GenAI isn't just built at the model level. It's shaped across the stack—from infrastructure to interface—and often captured at the moment of user impact. Great PMs don't just build smarter tools; they orchestrate systems where intelligence becomes actionable, trust becomes durable, and outcomes become scalable.

As you step into this space, think beyond individual components. Ask:

- How will your choices ripple across the stack?
- Where does your product sit in the GenAI value chain?
- And how will your design, pricing, and ecosystem strategy unlock both value and adoption?

The future of GenAI won't be shaped by those who chase hype but by those who understand where real value lies and have the product instincts to bring it to life. You're not just building with AI. You're building for a world that's being rewritten by it.

Chapter 8: Mastering GenAI Tools: Survive, Thrive, and Outpace the Market

AI may not replace PMs but PMs who master AI will replace those who don't. The pace of AI evolution isn't slowing down, it's accelerating. In the time it takes to scope a traditional feature, an AI-powered product team can test prototypes, iterate behaviors, and ship intelligent systems that adapt and learn in the wild. And it's not just about efficiency anymore, it's about survival. *Product managers who use AI do more, do it faster, and do it better than those who don't.*

In a world where product lifecycles are compressed, user expectations are dynamic, and AI models are improving monthly, tool fluency isn't optional—it's table stakes. The PMs who can build faster, think smarter, and orchestrate systems across infrastructure, model, logic & execution, hosting & serving, and other layers will define the next generation of product leadership.

Mastering GenAI tools isn't about chasing trends, it's about reshaping how you build, scale, and lead.

The best full-stack product makers don't just use GenAI tools, they leverage them. They treat every workflow, every bottleneck, and every opportunity as a chance to ask: *How can GenAI—or even agentic AI—help me work smarter, anticipate change faster, and expand what's possible?*

This chapter isn't about listing shiny new apps. It's about sharpening your instincts, expanding your capabilities, and equipping you to build the future today. Because full-stack product makers don't wait for change to happen. They build with it. In this chapter, you'll learn the following:

- The essential GenAI tool capabilities every PM must master
- How today's top tools supercharge each capability from ideation to optimization
- When to trust AI, when to override it, and how to build smarter, faster, and stronger
- How mastering GenAI tools shifts you from task executor to system orchestrator

By the end, you won't just use tools, you'll wield them with leverage, foresight, and adaptability, shaping products that thrive in the GenAI era.

How GenAI Tools Are Rewiring Product Workflows

AI isn't just accelerating individual tasks, it's reshaping the entire way full-stack product makers approach building, scaling, and evolving products. In the GenAI era, workflows that once moved predictably from one phase to another are collapsing, blurring, and accelerating. Understanding this new dynamic is key to thriving as a GenAI PM. GenAI isn't just speeding up product development, it's rewriting the entire playbook. To master GenAI tools, full-stack product makers must rewire how they think about workflows, strengths, and team dynamics.

From Linear to Parallel Workflows

Traditional product development followed a predictable, sequential flow: research led to design, design to development, development to testing, and finally, testing to launch. Each phase depended heavily on the previous one, creating bottlenecks that slowed progress whenever delays occurred upstream.

GenAI tools are disrupting this model entirely. They don't just accelerate individual steps, they allow multiple phases to unfold in parallel, enabling a more fluid and adaptive approach to product building. As shown in Figure 8.1, each phase of early product development—ideation, market research, concept testing, design, and MVP definition—can now be augmented by GenAI, enabling overlap and reducing the wait between stages.

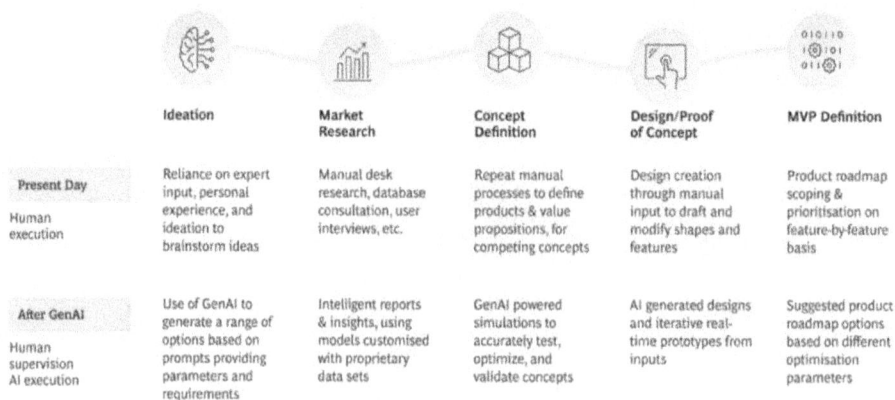

	Ideation	Market Research	Concept Definition	Design/Proof of Concept	MVP Definition
Present Day Human execution	Reliance on expert input, personal experience, and ideation to brainstorm ideas	Manual desk research, database consultation, user interviews, etc.	Repeat manual processes to define products & value propositions, for competing concepts	Design creation through manual input to draft and modify shapes and features	Product roadmap scoping & prioritisation on feature-by-feature basis
After GenAI Human supervision AI execution	Use of GenAI to generate a range of options based on prompts providing parameters and requirements	Intelligent reports & insights, using models customised with proprietary data sets	GenAI powered simulations to accurately test, optimize, and validate concepts	AI generated designs and iterative real-time prototypes from inputs	Suggested product roadmap options based on different optimisation parameters

Figure 8.1: How GenAI tools reshape early-stage product development (cropped from original). *Source: BCG. https://www.bcg.com/x/the-multiplier/genai-in-ecosystem-dynamics-optimizing-product-teams/*

Today, product teams can begin designing while AI-powered tools surface insights from user research in real time. For instance, large language models can analyze thousands of customer reviews to identify unmet needs—even as designers begin prototyping solutions. Figma AI can generate layout variations, while GitHub Copilot drafts backend logic in parallel. Meanwhile, AI simulators can test user flows dynamically as they're created, allowing for continuous feedback instead of waiting for traditional QA cycles.

This shift from sequential to simultaneous work fundamentally changes how full-stack PMs must think about speed and coordination. Instead of waiting for one team to hand off to the next, GenAI-powered workflows are multidirectional—blurring the boundaries between research, ideation, design, development, and testing. As a GenAI PM, your job is to architect workflows that are adaptive, responsive, and ready to evolve in real time.

Where GenAI Tools Excel—and Where They Don't

Despite their transformative power, GenAI tools aren't silver bullets. Strategic product managers must recognize where these tools provide leverage—and where human judgment still leads.

GenAI tools are especially effective in high-velocity, content-heavy domains. They excel at ideation and creativity, generating concepts, narratives, and design prototypes rapidly—often surfacing directions that would take teams hours to brainstorm manually. They're also strong at research synthesis, clustering user feedback, summarizing trends, and identifying latent patterns across large datasets.

In the early testing phase, GenAI can simulate user flows and predict friction points before any human even interacts with the interface. When it comes to drafting content—whether it's release notes, help center articles, or onboarding scripts—these tools shine. Their ability to generate, iterate, and adapt language makes them powerful allies for content-heavy product management work.

Figure 8.2 illustrates this clearly: GenAI tools drove nearly a 40% reduction in time spent on content-heavy tasks such as writing product requirements, FAQs, or backlogs. In contrast, content-light tasks like summarizing data or preparing one-pagers saw more modest gains, around 15%.

193

Average duration of task completion, minutes

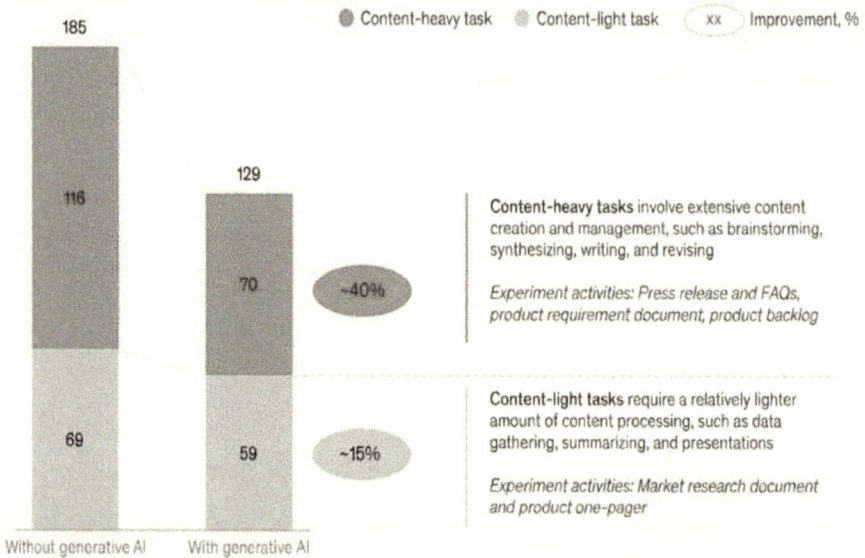

Content-heavy task   Content-light task   ⟨ xx ⟩ Improvement, %

Content-heavy tasks involve extensive content creation and management, such as brainstorming, synthesizing, writing, and revising

Experiment activities: Press release and FAQs, product requirement document, product backlog

Content-light tasks require a relatively lighter amount of content processing, such as data gathering, summarizing, and presentations

Experiment activities: Market research document and product one-pager

Without generative AI   With generative AI

Source: McKinsey gen AI and product management experiment (n = 40)

Figure 8.2: GenAI tools drive ~40% productivity gains in content-heavy product management tasks. Source: *McKinsey & Company. https://tinyurl.com/genai-productivity-gains*

That said, there are areas where human expertise remains critical—and likely will for the foreseeable future. Strategic planning still requires the kind of cross-functional trade-off thinking that tools can't yet replicate. Go-to-market execution depends on nuance, timing, and a deep understanding of competitive dynamics. Ethical decision-making—especially where trust, safety, or regulation is involved—demands a human-in-the-loop approach. And when systems fail in complex or unexpected ways, it takes human reasoning to identify root causes and course-correct.

By 2026, 75% of product managers will augment decision-making with AI, while prioritization, strategy, and ethics will remain firmly in human hands *(Gartner 2024)*—reinforcing the need to pair AI acceleration with human judgment. Mastering GenAI tools means knowing where they accelerate value—and where to steer manually using judgment, foresight, and experience.

Evolving Team Dynamics and Role Boundaries

GenAI isn't just changing what gets done, it's also changing who does it. Full-stack PMs are increasingly stepping into adjacent roles: drafting early design mockups with designers, setting up prompt flows and orchestrators with engineers, and running lightweight UX tests with researchers. This fluidity boosts speed and autonomy, but it also introduces tension.

A 2023 study found that PMs using GenAI tools not only experienced measurable productivity gains but also took on broader, cross-functional responsibilities compared to those in traditional workflows (McKinsey & Company 2024a).

New dynamics are emerging:

- *Expanded PM ownership:* Product managers are now orchestrating across model selection, infrastructure tuning, prompt engineering, and even UX validation.
- *Cross-functional overlap:* As AI flattens silos, designers and developers may feel their domains are being encroached upon by increasingly AI-empowered PMs.
- *Higher expectations for technical empathy:* PMs don't need to build models themselves, but they do need to understand trade-offs in latency, infrastructure cost, and model behavior to make smart product calls.

Full-stack product makers aren't replacing specialists, they're becoming high-leverage integrators. Thriving in this new dynamic means honoring the craft of each discipline while leading across evolving boundaries.

Building Faster, Smarter, and Stronger with GenAI Tools

Mastering GenAI tools isn't about chasing every new app—it's about building strength in the fundamentals that make great product managers effective. These tools won't replace your product judgment, but they can free up your time, sharpen your decisions, and help you focus on the parts of the job that truly require human insight.

Used well, GenAI tools allow you to move faster without losing clarity. They reduce bottlenecks, compress timelines, and help you move from messy inputs to useful outputs with far less friction. If you've ever spent days waiting on research synthesis, wrangling feedback across functions, or drafting a product spec from scratch—these tools can meaningfully change how you work.

In the next few sections, we'll walk through five areas where GenAI tools shift how PMs build—from prototyping and alignment to performance analysis and market sensing. Figure 8.3 gives a snapshot of how these tools are already reducing drag across common product functions, making teams quicker to respond, more coordinated, and better informed.

PM Function	Without GenAI	With GenAI Tools
Ideation and market research	Weeks of workshops and slow research cycles	Minutes to actionable insights (e.g., Perplexity AI Pro, ChatGPT Deep Research, Grok DeepSearch)
Prototyping flows	Weeks of specs	Hours to demo (e.g., Azure AI Foundry, GitHub Copilot Agent Mode, MCP Server, Microsoft Copilot Studio, Lovable, GitHub Spark, Bolt (bolt.new))
Shaping behavior	Engineering delays	Instant tweaks (e.g., Promptfoo)
Aligning squads	Endless meetings	Shared artifacts (e.g., Figma AI)
Analyzing performance	Manual logs	Real-time alerts (e.g., LangSmith)
Tracking trends	Research lag	Live insights (e.g., Perplexity)

Figure 8.3: Tool-driven PM efficiency.
© 2025 by the author of *Zero to GenAI Product Leader*.

Let's start with one of the clearest and most immediate ways GenAI can improve your workflow: how you discover, validate, and shape ideas.

Ideation and Market Research—From Guesswork to Insight

Getting to a strong idea used to take time. Now, it often starts with the right tools and a clear hypothesis. As a product manager working with GenAI, you still need to ground your thinking in user behavior, market gaps, and emerging patterns—but you don't always need to wait for lengthy workshops or slow research cycles to get there.

Generative tools don't replace thoughtful research or deeper exploration. But they do shorten the path from ambiguity to clarity, helping you form sharper hypotheses with more confidence, especially when time is tight or you're navigating new problem spaces. For example, **Perplexity Pro** can summarize content across forums, academic papers, and trusted media—useful for quickly exploring niche domains like "AI tools for adult learners with ADHD." **ChatGPT Deep Research**, launched in 2025, conducts multi-step exploration to produce citation-backed summaries, making it helpful for tasks like drafting market scans or understanding recent policy shifts. **Grok DeepSearch and DeeperSearch**, developed by xAI, combine real-time web signal extraction with deeper reasoning—often used to track fast-changing topics like model capabilities or user sentiment.

Some tools are better suited for brainstorming than for structured synthesis. **Claude 3.5**, for instance, supports open-ended ideation and tone exploration, particularly valuable when shaping narratives or early-stage product concepts. Meanwhile, tools like **Glean** and **OneAI** help make sense of large volumes of user feedback, clustering patterns from support tickets or app reviews without requiring custom NLP pipelines.

If you're monitoring developments across research or infrastructure, agent frameworks like **LangChain** or **Semantic Kernel** can automate routine tracking. Many PMs set up weekly digests that scan sources like GitHub, ArXiv, or model changelogs, summarizing key shifts relevant to their roadmap.

The goal isn't to replace your judgment—it's to reduce time spent gathering data, so you can spend more time making sense of it. In the GenAI era, ideation is less about isolated moments of inspiration and more about building repeatable ways to surface useful, timely ideas.

Prototyping: From Weeks to Hours

In fast-paced environments, ideas can't afford to sit still. The sooner you have something testable, the faster you can gather feedback and make better decisions. GenAI tools now make early prototyping faster and more accessible no matter your technical background.

Hugging Face Spaces, now with visual UI support, lets you spin up basic AI app demos in minutes—ideal for validating conversational flows or user journeys. Lovable, a browser-based prototype generator, enables PMs to design, test, and iterate agent interactions from plain-language descriptions and rough sketches before committing engineering time. Bolt (bolt.new) turns natural-language prompts—and even uploaded mockup images—into interactive web pages or app prototypes at the click of a button. GitHub Spark can spin up a full-stack micro-app from your written idea, so you can preview and refine a working prototype without writing code.

If you're comfortable with code or tackling more complex agents, GitHub Copilot's Agent Mode can generate and refine code in real time. Paired with MCP servers for model orchestration, it takes you from concept to a runnable agent in minutes. And even without a single line of code, Microsoft Copilot Studio lets you prototype workflows using natural-language prompts. With built-in connectors for M365 data and no-code orchestration blocks, you can simulate end-to-end assistant interactions and automate backend tasks independently. The goal isn't to get everything perfect on the first try, it's to test early, iterate fast, and reduce risk before the build begins. In the GenAI era, prototyping is less about polish and more about momentum.

Behavior Design and Testing: Iterate Without Bottlenecks

As GenAI products grow more complex, so does the challenge of keeping their behavior consistent. Models can drift—hallucinating facts, changing tone, or producing results that don't match intent. These shifts often happen without clear triggers, and if left unchecked, they can quietly undermine user trust.

In this environment, product managers can't afford to wait for full engineering cycles or model retraining to correct behavior. You need lightweight, reliable ways to test, tune, and adapt especially once a product is live.

Tools like Promptfoo, widely adopted by 2025, give PMs the ability to A/B test prompts, experiment with tone, and track behavioral consistency across different versions—without needing to write new code or block engineering teams. Grok 3 Playground, from xAI, supports structured testing of reasoning capabilities, helping catch overconfidence errors and prompt misalignments early.

These tools shift behavior design closer to the product team. What used to take weeks—flagging an issue, submitting an engineering ticket, waiting on a patch—can now be addressed in a day by iterating on the prompt or tuning the model's response format.

Consider how Notion AI handled this in 2024: when users flagged inaccuracies in meeting summaries, the team didn't retrain the model from scratch. Instead, they refined prompt instructions, added user feedback signals, and adjusted summarization logic—all without major engineering involvement. The result: improved output quality, faster response to feedback, and better user confidence.

Behavior design in GenAI isn't something you set once and forget. It's ongoing. Prompts act as prototypes, and frequent tuning helps keep the product aligned with expectations. Small changes, delivered regularly, often matter more than big overhauls delayed by long cycles.

Squad Alignment: Unify Teams with Clarity

In GenAI product development, speed can create more risk than reward if teams aren't aligned. Misunderstandings that might be minor in traditional cycles can snowball when AI behavior, infrastructure, and workflows evolve rapidly in parallel. That's why alignment needs to happen early and clearly. Long email threads, ad hoc notes, or verbal updates aren't enough when multiple teams are working on different parts of an intelligent system.

Tools like Microsoft Copilot for Teams help reduce friction by generating structured meeting notes, task breakdowns, and decision summaries automatically. Instead of relying on someone to transcribe and share takeaways, the team can start from a shared set of AI-generated artifacts—often within minutes of the discussion ending.

Visual tools like Miro, now integrated with Miro AI, add another layer of clarity. Teams can map interactions between users, agents, APIs, and backend logic in real time—highlighting where breakdowns might occur before any development begins. These shared canvases make it easier to spot misaligned assumptions or missing orchestration steps early, when fixes are still easy to make.

The goal isn't to replace meetings, but to make collaboration more focused and visual. Rather than explaining how something *should* work, product teams can simulate and inspect it together, creating faster alignment and fewer surprises downstream. In high-velocity GenAI projects, clarity isn't just about better communication—it's about using the right tools to show, not just tell. PMs who prioritize early simulation and shared documentation often avoid the rework that slows everyone else down.

Performance Monitoring: Catch Errors Early

In GenAI products, problems don't always show up right away. Sometimes an agent gives an off response. Sometimes a prompt stops working the way it used to. These issues like hallucinations, context loss, or tool failures can quietly affect the user experience without anyone noticing at first.

If you don't catch them early, they build up. They create frustration for users, damage trust, and can even lead to compliance risks. That's why monitoring these systems isn't just an engineering task. As a product manager, staying close to how the system behaves is part of your job.

Fortunately, the tools to help you do that are getting better. LangSmith lets you follow how prompts and tools are triggered in your workflows, so you can see where things might be breaking down. OpenAI's Evals now support multimodal content, giving you a way to track how reliably your product is performing across text, images, or audio. Tools like Arize AI help you monitor model accuracy, user satisfaction, and performance shifts over time. And PromptLayer gives you an easy way to compare how prompt changes are affecting responses across experiments.

You don't have to build all this from scratch. But you do need a system that helps you notice issues early, before they reach users. A simple threshold, like flagging when hallucination rates go over 5%, can help you spot and fix problems before they get worse.

Great GenAI products aren't just fast, they're dependable. And the teams that build them know that catching small issues early is what keeps trust high over time.

Trend Tracking: Stay Ahead of the Curve

The pace of change in GenAI is unusually fast—new models, capabilities, frameworks, and user behaviors emerge regularly. What was cutting-edge last month may feel dated today. In more traditional domains, product teams could afford to adopt new trends slowly. In GenAI, delayed awareness can lead to strategic missteps.

While ideation tools help clarify what users need now, trend-tracking helps you prepare for what they might expect next. This isn't just about staying current. It's about staying relevant.

Many product managers still rely on periodic analyst updates or industry reports. But in fast-moving spaces like GenAI, those cycles aren't fast enough. Leading PMs build lightweight systems, often powered by GenAI tools, that keep them informed on a weekly or even daily basis.

Tools and techniques vary:

- Summarization bots built on platforms like ArXiv or Hugging Face help surface the key takeaways from dense research papers without requiring a deep technical background.

- Custom GPTs and market intelligence agents can be configured to monitor updates from specific labs or competitors, surfacing relevant changes without overwhelming detail.
- Open-source trackers and feeds like model leaderboards, changelogs, or trusted voices on platforms like X offer faster visibility into major developments.

Some product managers go further and use orchestration frameworks like LangChain, Semantic Kernel, or Microsoft Copilot Studio to automate trend tracking. For example, a simple setup might monitor GitHub repositories, summarize updates from AI forums, and deliver a weekly digest highlighting developments that could influence your roadmap.

You don't need to build elaborate systems to benefit. The key is consistency—having a clear mechanism for staying informed, without it becoming a burden.

You're not expected to be an AI researcher. But you are expected to sense shifts early, filter what matters, and translate those signals into product direction. In fast-changing spaces, that kind of strategic awareness becomes a baseline expectation—not a bonus skill.

Advanced skill: building your own research agents

Leading full-stack PMs don't wait for information—they architect how they consume it. Today, they can build lightweight research agents—using platforms like Azure AI Foundry or Microsoft Copilot Studio, Semantic Kernel, or LangChain—to stay ahead of ecosystem shifts.

This doesn't require deep engineering expertise or a custom tech stack. Most agents begin with a simple setup: monitoring a few key sources like GitHub, Hugging Face, ArXiv, or news feeds. One component retrieves updates. Another summarizes key developments. A third helps prioritize what matters most for your product area. Some teams go a step further and route this into a Monday-morning digest, offering just-in-time awareness to inform roadmap conversations or sprint planning.

What starts as a side experiment often becomes a quiet but durable habit. Instead of reacting to trends when they become mainstream, you develop the infrastructure to spot them earlier, evaluate their relevance, and act with clarity.

This approach also benefits teams. When multiple stakeholders share the same signal feed, conversations move faster and decisions are more grounded. No one is chasing information after it's already shaped the market.

In a space that changes weekly, building your own research agent isn't about chasing trends for the sake of novelty. It's about staying focused on the right signals and structuring your learning flow in a way that keeps you ahead—quietly, consistently, and with intention.

Considerations for Selecting GenAI Tools

The number of GenAI tools available today can feel overwhelming. It's easy to get caught in one of two extremes: chasing the latest releases without clear purpose, or sticking only to your company's default tools even when better options exist.

Strong product managers avoid both. They treat tool selection not as a reaction to hype, but as a deliberate choice grounded in what their workflows demand. The right tool isn't just the one with the most features, it's the one that meaningfully reduces friction in how you discover insights, test ideas, and build with speed and confidence.

The following principles can help you choose GenAI tools that align with how you work and where you're trying to go.

Start with your friction points, not the tool features

The best starting point isn't the tool, it's the bottleneck. Instead of asking "What can this tool do?", ask "Where am I currently stuck?"

For example, if synthesizing research is taking too long, tools like Perplexity or ChatGPT with browsing mode may help you extract useful signals faster. If validating prototypes early is a challenge, platforms like Hugging Face Spaces, Lovable or GitHub Copilot Agent Mode can support quick iterations before engineering is involved.

If you're seeing behavior inconsistencies in model output, tools like Promptfoo or LangSmith are designed to help test and track those issues without relying heavily on engineering.

If the challenge is team alignment, visual platforms such as Miro or Figma—especially with built-in simulation features—can help teams stay on the same page earlier in the design and development process.

For monitoring issues like model drift or latency gaps, observability tools like Arize and LangSmith are built to surface silent failures before they become user-facing problems.

The core question isn't "What's trending?" It's "Where am I losing time, clarity, or trust?" Let that guide your tool selection not a feature list.

Match the tool to the stage of the GenAI product lifecycle

Not every GenAI tool is designed for the same moment in the product lifecycle. Some help you explore and validate ideas early. Others are built to monitor behavior, test quality, or improve how systems scale.

Before choosing a tool, step back and ask: *Which stage of my product am I optimizing right now?* Are you prototyping, shaping behavior, running tests, or preparing to scale?

In the **ideation and prototyping** phase, you want to spin up ideas fast without waiting on engineering. No-code and low-code environments like *Figma with GenAI*, *Miro AI*, *Lovable*, and *Bolt.new* let you simulate user flows, experiment with prompt designs, and iterate in real time.

Once you've sketched out a concept, it's time for **behavior shaping and iteration**. Platforms such as *Promptfoo* and *Grok 3 Playground* empower you to fine-tune prompts, evaluate model responses, and gradually steer your system toward the right behaviours.

When your prototype moves into production-like conditions, **testing and monitoring** become critical. Tools like *LangSmith*, *Arize AI*, and *PromptLayer* provide drift detection, output-quality dashboards, and alerting—so you catch problems before they impact users.

Finally, as you **prepare to launch and scale**, observability solutions (*Arize AI* or *LangSmith*) give you end-to-end telemetry, helping you understand system health, user adoption, and performance across growing workloads.

The key is not to use every tool, but to **match the right tool** to the moment in your GenAI product lifecycle. Figure 8.4 shows how tool selection can align with different stages in the GenAI product lifecycle. It's not exhaustive, but it can help you think more systematically about what you need and when.

Lifecycle Stage	PM Tool Focus	Example Tools
Ideation & prototyping	Flow simulation, prompt design	Figma with GenAI, Miro AI, Lovable, Bolt.new
Behavior shaping & iteration	Prompt tuning, behavior evaluation	Promptfoo, Grok 3 Playground
Testing & monitoring	Drift detection, evaluation dashboards	LangSmith, Arize AI, PromptLayer
Launch & scaling	Observability, telemetry-driven iteration	Arize AI, LangSmith

Figure 8.4: Tools across the GenAI product lifecycle

Prioritize tools that make you *independent*, not dependent

The best GenAI tools don't create new bottlenecks, they remove them. If a tool requires an engineer to set it up, troubleshoot it, or interpret its results every time you use it, it's likely not the right fit for a fast-moving product team. As a PM, you should look for tools that put more of the process in your own hands.

Whether you're designing a prompt test, interpreting telemetry, or simulating agent behavior, the goal is to move without being blocked. Tools that let you run small experiments, visualize system behavior, or act on data directly—without needing custom setup—tend to scale better with the pace of your work. One useful way to evaluate a tool is by asking, "Does this help me move independently, or will I still be waiting for someone else to unblock me?"

Choose tools that integrate across your stack

A tool can be powerful on its own and still create friction if it doesn't fit with the rest of your workflow. Many PMs find that standalone tools become a source of overhead, especially when they require exporting results manually or duplicating work across systems.

When evaluating a new tool, consider how well it connects to what your team already uses. Tools that can plug into systems like GitHub, Azure AI, Jira, Slack, LangChain, or VS Code tend to reduce context-switching and improve follow-through.

They also make it easier to share artifacts with engineering or design partners in a form that's ready for action. The simplest way to test fit is to ask: "Will this tool amplify the systems we already rely on—or will it sit in a silo that creates extra steps?"

Check organizational compliance and security requirements—early

Before piloting or adopting any GenAI tool, it's important to understand how it aligns with your company's compliance and security requirements. This isn't just a legal concern; it can save weeks of back-and-forth later.

Make sure you know where the tool stores prompt and response data, how it handles personally identifiable information, and whether it meets your company's standards for encryption and access control. Some tools are designed with enterprise use in mind and offer clear documentation and controls. Others may be more experimental or consumer-focused and carry risk when used with sensitive inputs.

If a tool hasn't already been reviewed by your security team, it's worth checking whether vendor approval is required before putting it into production use. The key question to ask early is: "Could this tool expose user data, model behavior, or company IP in ways that violate internal policies?"

Understanding the boundaries before you start integrating can help you move faster later and avoid needing to unwind decisions after the fact.

Optimize for speed-to-feedback without compromising governance.

Fast iteration helps you move quickly, but if it comes at the cost of compliance or security, it can set you back in far more serious ways. Whether you're prototyping prompts, testing workflows, or evaluating new tooling, always confirm that the systems you're using meet your organization's data and privacy standards.

A good GenAI tool should reduce delays but not by sidestepping the guardrails your company has put in place. In product work, speed helps you stay ahead, but trust keeps you in the market. Choose tools that support both.

Prioritize learning curve vs. payoff

Some tools offer more flexibility or customization but take time to learn. Others provide quick wins but are limited in how far they'll scale with your work. Neither is better by default; it depends on what you're trying to accomplish and how often you'll need the tool.

For example, LangSmith gives you deep control over prompt testing and orchestration workflows, but it takes effort to learn and configure. ChatGPT plug-ins, on the other hand, are easier to use out of the box but may not support the complexity you need over time.

The key is to think in terms of leverage. Will this tool serve you across multiple projects? Does it reduce bottlenecks or improve quality in ways that compound over time? If so, it may be worth the upfront investment.

Tool selection isn't about covering every option, it's about making a few well-informed choices that improve how you work and build. As the ecosystem continues to evolve, the real advantage comes from going deep on the tools that matter, not trying to keep up with all of them.

Squads of Tomorrow: Human and Agent PMs in Harmony

The future of product management isn't human-led or AI-driven, it's a dynamic partnership. Human PMs and agent PMs will co-lead GenAI squads, blending human creativity with AI precision.

Full-stack product managers won't just orchestrate researchers, designers, and engineers. They'll lead fleets of AI agents that amplify their impact by doing the following:

- Scanning user trends and telemetry 24/7
- Auto-optimizing prompts to maintain model performance
- Synthesizing feedback to propose experiments
- Drafting prototypes and simulating user journeys

Agent PMs will autonomously handle operational tasks—monitoring, testing, and iterating—freeing human PMs to focus on vision, empathy, and ethical stewardship.

See Figure 8.5 below for a detailed breakdown of their complementary roles.

Responsibility	Human PM	Agent PM
Define product vision, strategy, and success metrics	✓	✗
Monitor telemetry for trends, drift, or anomalies	✓ Oversight	✓ Execution
Propose prompt refinements or UX iterations	✓ Review & prioritize	✓ Auto-propose or test
Draft user flows or agent orchestration diagrams	✓ Define scenarios	✓ Generate drafts
Identify risks, ethical concerns, or trust issues	✓	✗
Align cross-functional human teams	✓	✗
Plan and execute A/B tests or small experiments	✓ Supervise results	✓ Execute
Craft narratives, roadmaps, and executive comms	✓	✗

Figure 8.5: Division of labor: human PM vs. agent PM.
© 2025 by the author of Zero to GenAI Product Leader.

The Evolving Role of Human PM

As agents rise to take on operational load, the role of the human PM doesn't shrink—it sharpens. You're no longer the center of every task. You're the conductor of an intelligent, evolving system.

Next, we'll cover how your role transforms in this new era.

Vision architects

You will define the *why* behind every initiative. While agent PMs can optimize prompts or ship experiments, they can't set the North Star. That's your job.

You'll map out where the product needs to go and ensure every human and agent on the squad is aligned. It's about crafting a compelling product narrative and continuously asking: *"What future are we building and for whom?"*

Ethical guardians

GenAI introduces new forms of risk: hallucinations, bias, over-automation. You are the one who must pause and ask: *"Just because we can, should we?"* You'll integrate tools that audit agent decisions, surface edge cases, and enforce responsible AI guardrails. In this world, ethics isn't compliance—it's design. You lead with values.

System orchestrators

You're no longer just coordinating humans, you're choreographing complex systems of APIs, agents, and humans working in sync. That means designing workflows where feedback loops are tight, responsibilities are clear, and agents don't just act—they collaborate. You'll work with tools like Langchain, Semantic Kernel, and orchestration frameworks to ensure every agent operates in harmony—not isolation.

Outcome optimizers

You'll use agent-generated insights not just to react—but to *learn*. Did a summarization agent start missing key points? You'll adapt the prompt or retrain. Did feedback loops reveal user confusion? You'll rework flows. Your job is to connect the dots between what agents see and what users need, turning insight into iteration, faster than ever.

This is not about giving up control. It's about reclaiming your time and refocusing your energy—on strategy, creativity, and building products that deserve to exist.

Agent PMs won't replace you. They'll *amplify your effectiveness*—by surfacing patterns, handling repeatable tasks, and offering real-time eyes on everything from prompt drift to user friction. But the soul of the product? The vision, the values, the judgment? That's still all you.

A day in the life: human + agent PMs in flow

It's 8:45 AM in Seattle.

Maya, a GenAI product manager, sits down with her morning tea—not to check Slack or emails, but to review her agent-generated daily briefing. Her master orchestration agent, plugged into LangSmith and Arize, has already done the following:

- Flagged a spike in hallucination rates on one of the new summarization flows
- Compiled the three most insightful user feedback threads from GitHub and Discord
- Suggested an A/B prompt tweak based on past fine-tuning data
- Summarized key trends from developer forums that could inform next quarter's roadmap

By 9:00 AM, Maya hasn't just caught up, she's ahead.

At 10:15 AM, she hops on a call with legal and design to review the new ethical review dashboard her compliance agent flagged. A spike in user queries around medical advice prompted the system to surface safety concerns. Maya asks the ethics agent to simulate different prompt restrictions and their downstream impact on UX. It does this in minutes.

By noon, her experiment agent has already spun up two alternate flows and deployed them to a 5% cohort. Results will trickle in by end of day—no JIRA tickets, no bottlenecks.

That afternoon, Maya meets with her engineering lead. They don't talk about operational issues—agents already surfaced those and created GitHub issues with proposed remediations. They discuss *vision*: How can they push the product toward more proactive assistance without crossing the line into user discomfort?

At 5:00 PM, Maya wraps.
Her agents will keep scanning signals and telemetry overnight. She ends her day the way she started it—not reacting, but *steering*.

Here's what you just saw:

- Maya didn't write code.
- She didn't wait on dashboards or analysts.
- She didn't spend hours in backlogs.

She led with judgment, acted with insight, and collaborated with agents that made her faster, more focused, and more impactful. This is not science fiction. This is the new reality of GenAI product management.

And if you're an aspiring PM? Start now. Start small. Start by mastering the tools because soon, your agents will be looking to you for direction.

In the GenAI world, teams don't just move faster. They think faster, test faster, and learn faster. Your squad won't be limited by headcount—it will be defined by how well you orchestrate human ingenuity and AI precision. The best PMs won't ask, "Can I get more resources?" They'll ask, "Which agent can I spin up to amplify this task?"

If you're just getting started, here's the truth: Mastering GenAI tools is no longer a bonus—it's your new baseline.

To thrive, you'll need to do the following:

- Build agile, lightweight squads of agents that evolve with your product

209

- Use tools like LangChain, Semantic Kernel, and Azure AI Agent Services to compose agent workflows
- Lead not just people, but ecosystems—where every agent, every insight, and every iteration loops back into smarter decisions

The game has changed. Product management isn't about wrangling tasks—it's about designing intelligent systems that can self-improve and scale trust. The revolution has already begun. And if you're reading this, you're not late—you're right on time.

Conclusion: You Don't Just Use Tools—You Lead with Them

By now, one thing should be clear: mastering GenAI tools isn't a technical upgrade. It's a leadership transformation.

This chapter wasn't about learning shortcuts. It was about building leverage—about using tools not to work more, but to work differently. More fluidly. More intelligently.

You've seen how GenAI tools can compress timelines, expose drift before it causes damage, surface user insights before your competitors do, and simulate products before you've even booked your first sprint planning.

You've also seen where tools can't go alone. They won't define your vision. They won't make ethical trade-offs. They won't know when *not* to automate. That's where you come in. Because in this new world, tools don't replace product managers, they extend them.

The best GenAI PMs know when to delegate to agents, when to intervene with judgment, and when to rewire the system itself. So, as the ecosystem accelerates, ask yourself the following questions:

- Where am I still waiting on others when I could act with a tool?
- Which workflow is lagging because I haven't automated it yet?
- What decision could I make faster, if only I asked an agent first?

The next generation of great products won't come from teams that work harder. They'll come from PMs who build smarter, by wielding the right tools at the right time to shape the right outcomes.

You don't need to master every model, every platform, or every agent framework overnight. But you do need to start. Today. Because in the GenAI era, leadership isn't just about having the answers. It's about building the system that finds them.

The future is already forming. And the tools to shape it are right in front of you. Pick one. Spin up your first agent. Redesign a workflow. Build your edge. This is your moment to stop chasing change and start driving it.

Chapter 9: Making Better Decisions as a GenAI Product Manager

As a GenAI product manager, you navigate a landscape filled with uncertainty. Perfect data, clear roadmaps, or foolproof playbooks? They're rare in this field. Instead, you're tasked with making high-stakes decisions about cutting-edge technologies, innovative user experiences, and risks that haven't been fully mapped—all under tight deadlines. In my years as a product manager, I've learned that the best PMs don't just react to this ambiguity; they shape it. Strategic decision-making is your greatest asset, and it's what separates good PMs from great ones.

In Chapter 5, we walked through the hands-on process of creating a GenAI solution—framing user problems, choosing interaction patterns, and scaling systems. Tools like the *four-lens framework, introduced in Chapter 5* guided you through execution-focused decisions, such as selecting use cases or prioritizing features. That chapter was your blueprint for getting things done.

Here in Chapter 9, we shift to the bigger picture: strategic decision-making. This is about tackling the tough questions that define your product's path: Is this idea worth your team's time? Should you build, buy, or integrate a solution? What should you prioritize today versus next quarter? While Chapter 5 equipped you to build with precision, this chapter is your compass, offering advanced frameworks tailored to the complexities of GenAI. I've seen teams falter by diving in without asking these questions first—don't let that be you.

In this chapter, my goal is to give you clear, practical tools to analyze trade-offs and move forward with confidence, strengthening both your product and your credibility. We'll explore **seven critical decision moments**, each paired with frameworks designed specifically for you as a GenAI product manager.

Decision 1: Is GenAI Right for Your Project?

As a GenAI product manager, your first and most pivotal decision is whether GenAI is the right technology for the challenge you're tackling. It's tempting to apply AI to every problem—after all, the buzz around GenAI is hard to ignore. But I've learned, sometimes the hard way, that not every problem benefits from a generative approach. Some are better solved with simpler tools, while others that seem promising can crumble under scrutiny due to data gaps, excessive complexity, or low business value. In my career, I've seen teams rush into GenAI projects, only to find that a basic algorithm would have sufficed or that the promised value didn't materialize. This decision is the foundation of your project, everything else, from model selection to UX design, hinges on getting it right.

In Chapter 5, we used the *four-lens* framework to refine a GenAI use case once you've committed to a project, ensuring a user-centric and feasible solution. Here, we step back to ask a more fundamental question: Should you pursue GenAI at all? These are the questions I ask myself before greenlighting any GenAI initiative, and they'll guide you to make strategic choices that avoid costly missteps.

Key Questions

Does your problem truly require GenAI?

GenAI shines in tasks demanding creativity, natural language processing, or complex pattern recognition—like generating content, translating languages, or suggesting code. But for straightforward problems, such as data sorting or rule-based automation, simpler tools like algorithms or traditional machine learning are often more efficient and cost-effective. Ask yourself: Does the complexity of your problem justify GenAI's power, or could a leaner approach deliver the same results with less effort?

Is the task generative, predictive, or reasoning-based?

To confirm GenAI's fit, clarify your task's nature. GenAI excels in generative tasks, such as creating text, images, or code, where its ability to produce novel outputs is a strength. It can also support predictive tasks, like forecasting trends, or reasoning-based tasks, where it navigates ambiguity to simulate human-like decisions. For example, if you're predicting inventory needs, a dedicated predictive model might be better. But if you're drafting personalized emails, GenAI's generative capabilities are ideal. Understanding whether your task is generative, predictive, or reasoning-based ensures you're leveraging the right technology.

Will It deliver significant value?

Even if GenAI is technically viable, you must verify that it creates meaningful value for your users or business. Will it enhance user experiences, streamline operations, reduce costs, or unlock new opportunities? I've seen technically impressive projects fail to gain traction because they didn't deliver enough impact. Evaluate the value carefully: If the benefits are marginal or achievable with less effort, GenAI may not be worth the investment.

Tools

AI Relevance Canvas

The *AI Relevance Canvas* is a powerful tool to stress-test whether GenAI is a smart fit or a distraction.

Unlike Chapter 5's *four* lenses, which refine a GenAI project's scope, this canvas, shown in Figure 9.1, helps you decide if GenAI is the right technology at all. It outlines five dimensions to map your problem and assess its suitability for GenAI, ensuring you consider critical factors like task type, data needs, and risks.

Dimension	What to Ask	Why It Matters
Problem Type	Is the task generative, creative, or adaptive?	GenAI shines at open-ended, unstructured tasks—not CRUD apps.
Input & Output Structure	Are inputs messy? Are outputs open to variation?	GenAI works well when rules are fuzzy and answers aren't fixed.
Need for Learning	Will the system benefit from learning or adapting over time?	Static rules may be better if the task rarely changes.
Data Availability	Do we have relevant data—or examples—to support GenAI?	Garbage in, garbage out. No training signal = poor results.
Risk Level	Will hallucinations or ambiguity cause harm?	In regulated or high-stakes domains, GenAI may be risky.

Figure 9.1: AI relevance canvas for evaluating GenAI suitability.
© 2025 by the author of *Zero to GenAI Product Leader*.

Value/Effort Matrix

The *Value/Effort Matrix* (see Figure 9.2) helps you assess whether a GenAI project is worth pursuing by mapping its expected impact (value) against the resources and complexity it will require (effort).

Value might include improvements to user experience, business efficiency, or strategic differentiation. Effort includes engineering complexity, model readiness, compliance overhead, and team ramp-up time.

Used early in the scoping process, this tool helps teams align quickly on which initiatives to prioritize, which to delay, and which may not be worth pursuing at all.

The matrix includes four quadrants:

- **Quick Win**: High value, low effort. These are your top priorities—projects you can deliver quickly with tangible benefits.
- **Strategic Investment**: High value, high effort. Worth pursuing, but often requires sequencing, phased rollout, or cross-team alignment.
- **Reconsider**: Low value, low effort. These projects may be useful for learning or prototyping but aren't likely to justify sustained investment.
- **Drop or Park**: Low value, high effort. These are often distractions. They tend to drain resources without delivering enough return.

Figure 9.2: Value/Effort Matrix

Even a quick team exercise using this matrix can clarify trade-offs and reveal mismatches between perceived impact and true complexity. It's not a decision-making tool on its own but it makes conversations more grounded and actionable.

And it's not just useful when you're evaluating whether to pursue GenAI at all. As your portfolio grows, the same matrix becomes a practical way to prioritize across competing GenAI features or initiatives. Whether you're choosing between a quick enhancement to boost adoption or a long-term bet on agentic workflows, the framework helps surface what matters most: impact, effort, and timing. What starts as a scoping tool can evolve into a reliable compass for prioritizing smartly, even as your product matures.

Chapter 5 guided you through shaping a GenAI project once you've committed to building. This section takes a step back to help you validate whether that commitment should happen in the first place. By combining the *AI Relevance Canvas* with the Value/Effort Matrix, you give your team a sharper filter—ensuring you're solving the right problems, with the right methods, for the right reasons.

This kind of upfront validation isn't theoretical; it plays out in real product decisions. One of the clearest examples I've seen is GitHub Copilot, launched in 2021 through a partnership between GitHub and OpenAI. A GenAI-powered coding assistant I rely on myself, Copilot shows how the frameworks above can guide smart bets. Does coding require GenAI? In my experience, yes—writing code demands creativity and contextual understanding, where GenAI excels (problem type: generative, creative).

Static templates can't adapt to diverse coding styles or project needs (input & output structure: messy inputs, variable outputs). Copilot learns from your coding patterns, improving suggestions over time (need for learning: high). GitHub leveraged vast code repositories to train it, ensuring robust data (data availability: strong). While errors in code suggestions carry some risk, they're low-stakes in development environments with human oversight (risk level: manageable).

The Value/Effort Matrix shows significant effort but immense value—a GitHub study conducted in collaboration with Accenture found that developers using Copilot completed tasks up to 55% faster in enterprise settings (GitHub 2024). And it's not just useful for engineers—recent versions of Copilot, including Agent Mode, help product managers prototype flows, review logic, and collaborate faster with engineering, making it a valuable tool across the product lifecycle. Copilot's success underscores why this kind of upfront evaluation matters.

With GenAI confirmed as the right fit, the next step is deciding what kind of system to build, covered in the following section.

Decision 2: What Kind of GenAI System Should You Build?

Once you've confirmed that GenAI is the right fit for your project, the next pivotal decision is choosing the system architecture that best suits your needs. This choice defines how your product functions, performs, and delivers value, and it's where strategic thinking can set you apart.

In my years as a product manager, I've seen teams falter by chasing the latest GenAI architecture without aligning it with their project's goals. I once advised a team that opted for a complex agentic system for a simple query task, only to spend months untangling unnecessary complexity. That experience taught me to approach architecture decisions with clarity and purpose. This section provides tools to help you select the right GenAI system type, ensuring it matches your task's requirements and sets your project up for success.

In Chapter 5, we explored system design patterns like prompt, RAG, and agentic RAG, guiding you through execution-focused choices within a GenAI project. Here, we take a strategic perspective, focusing on the high-level question of which architecture to pursue before diving into implementation. Use these questions as a blueprint for shaping GenAI system behavior that balances user trust, autonomy, and real-world reliability.

Key Questions

What system architecture is best suited for your task?

GenAI systems vary widely, from retrieval-based setups like retrieval-augmented generation to fine-tuned models or autonomous agentic systems. Each has distinct strengths:

- RAG: Excels at grounding responses in specific data, ideal for knowledge-heavy tasks like searching a legal document database.
- Fine-tuning: Optimizes for task-specific performance, perfect for applications like tone-specific copywriting or classification.
- Agentic systems: Handle multi-step workflows, such as automating task recommendations or orchestrating complex processes. Ask yourself: Does your task require factual retrieval, personalized behavior, or dynamic reasoning over multiple steps? The answer will point you to the architecture that aligns with your project's complexity and goals.

What are your priorities for quality, cost, and speed?

Every architecture comes with trade-offs. A fine-tuned model might deliver high-quality outputs but require significant investment. RAG could be cost-effective but slower for complex queries. Agentic systems offer automation but demand robust infrastructure.

Reflect on your priorities: Do you need top-tier quality, even if it's costly? Is speed to market critical, even if it means lower customization? Understanding these trade-offs helps you choose an architecture that fits your constraints.

Will It run as a standalone tool or a collaborative system?

Consider how your system will interact with users or other workflows. Will it function as a standalone tool, like a knowledge base search, or as a collaborative system, like a copilot embedded in a broader platform? For example, a standalone RAG system might answer queries independently, while an agentic system could integrate with a project management tool to suggest actions. Reflect on your use case: Is your system a self-contained solution, or does it need to enhance an existing workflow?

Tools

RFA Lens

The *RFA lens* is a decision tool you can use to choose between "retrieval" (RAG), "fine-tuning," or "agentic" architectures. As shown in Figure 9.3, it maps your task's requirements—data grounding, task specificity, or workflow automation—against each option's strengths. You don't always have to pick just one; many advanced systems combine them. But starting with Figure 9.3 clarifies the primary architecture to build around.

Option	When to Use	Common Use Case
R – Retrieve (RAG)	When knowledge is static but broad, and context is key	Legal document Q&A, customer support knowledge bases
F – Fine-tune	When task behavior must be precise and replicable	Classification, summarization, tone-specific copywriting
A – Agentic	When the system needs to reason, plan, or take action over time	Copilots, workflow automation, decision support systems

Figure 9.3: RFA lens for GenAI architecture decisions.

The Value/Effort Matrix (Figure 9.2), introduced in Decision 1, is also useful here. Use it to prioritize architecture choices. A high-value, low-effort option like RAG for a data-heavy task is a quick win, while a high-effort, high-value choice like fine-tuning might be a strategic investment. Low-value options should be reconsidered or dropped.

Data Decisions: A Critical Checkpoint

Choosing an architecture is only half the battle, your system's success hinges on the data behind it. I've seen projects derail because teams locked in an architecture without checking their data readiness, like picking an engine without ensuring

218

there's fuel available to run it. This checkpoint ensures that your choice is viable, saving you from costly pivots later.

The following questions are good to ask:

- Do you have the right data to support your system design choice?
  - For RAG: Is your knowledge base fresh, searchable, and reliable? Stale or disorganized data will lead to irrelevant responses.
  - For fine-tuning: Do you have task-specific, high-quality labeled data? Poor labeling can corrupt your model's performance.
  - For agentic systems: Do you have structured data to fuel actions and memory? Without clear data, agents can't reason or act effectively.
- What are the risks?
  - Stale or missing data ruins retrieval systems.
  - Poor labeling undermines fine-tuned models.
  - Privacy violations or unclear data ownership can lead to IP or legal issues.

Choosing a system without evaluating your data readiness is a recipe for failure. By using the *RFA lens* to select an architecture—and then checkpointing against your data realities—you ensure your decision is both strategic and feasible. Smart PMs validate their choices early to avoid costly pivots.

Applying tools like the *RFA lens* and a structured data assessment helps you avoid the trap of over-engineering or selecting an ill-suited architecture—a mistake I've seen derail even the most promising projects. This decision ensures your system is grounded and viable, paving the way for the sourcing trade-offs we'll explore next.

Consider Microsoft 365 Copilot. It integrates GenAI into everyday workflows across Word, Outlook, Excel, and Teams using a layered architecture that blends retrieval, fine-tuning, and agentic behavior. Its retrieval layer gathers information from a user's documents, emails, and Teams messages—so when someone asks Copilot to summarize the last three meetings with a client, it pulls relevant content and generates a tailored response. On top of that, Microsoft fine-tuned smaller internal models to adapt outputs for tone, formatting, and enterprise consistency—for instance, ensuring that suggested emails align with company style guides. At its most advanced layer, Copilot exhibits agentic behavior: it not only responds but also proposes next steps like scheduling a meeting or assigning follow-ups in Teams.

This system wasn't built on a single architectural bet. It reflects a hybrid strategy—where large models like GPT-4 are orchestrated alongside task-specific models, all governed by enterprise-grade guardrails. The result is a system that delivers the intelligence of foundation models while remaining context-aware, secure, and practical within real enterprise workflows (Microsoft 2023).

Decision 3: Should You Build, Buy, or Create a Hybrid?

After selecting your GenAI system's architecture, the next critical decision is how to source it: should you build it in-house, buy a pre-built solution, or adopt a hybrid approach that blends custom and off-the-shelf components? This choice shapes your project's timeline, budget, scalability, and adaptability, making it a cornerstone of your long-term strategy. In my experience as a product manager, getting this decision right often makes the difference between shipping fast with confidence and getting stuck in cycles of regret or rework.

In Chapter 5, we explored integrating APIs and off-the-shelf tools during execution. Here, we zoom out to ask a more strategic question: how should you acquire the system's core capabilities in the first place? You'll evaluate this through eight criteria grouped into three themes: control and ownership, feasibility and execution, and risk and strategic alignment. These are the dimensions I revisit with every GenAI project—especially those involving agentic behavior or enterprise-grade orchestration.

So how should you decide? Start with the sourcing paths outlined in Figure 9.4. Each one has distinct trade-offs, and your best option depends on the constraints and goals of your product.

Build:
Build means developing your GenAI system entirely in-house. You control the data flows, system behavior, orchestration logic, and IP. This path is often chosen when you need deep vertical integration, custom behavior, or strict compliance—especially for agentic systems in regulated industries.

Buy:
Buy refers to licensing pre-built models, APIs, or complete GenAI platforms from vendors. It's the fastest way to get started and often the most cost-efficient, particularly when validating use cases, enhancing non-core workflows, or embedding

copilots. However, you trade control for convenience, and vendor limitations may slow future scaling.

Hybrid:

Hybrid blends the two approaches. You might combine off-the-shelf models or APIs with your own custom components—such as wrappers, toolchains, datasets, or orchestration logic. This setup is especially common in agentic systems, where teams use hosted frameworks (like LangChain, or Semantic Kernel) to speed up development, but retain control over memory, tool access, policies, or evaluation. Hybrid gives you flexibility where it matters, while keeping time-to-market and engineering effort within reach.

Option	Pros	Cons	When It Makes Sense
Build	High control, custom IP	Slow, costly, resource-heavy	Unique data, strict compliance needs, deep vertical integration, full agent customization
Buy	Fast, low-cost, simple to deploy	Less control, vendor lock-in or gaps	Quick market validation, non-core enhancement, early-stage experiments
Hybrid	Balanced speed and flexibility; component-level control	More complex to integrate and maintain	Agentic systems needing orchestration + custom logic; fast entry with long-term adaptability

Figure 9.4: Build, buy, or hybrid options for GenAI system sourcing

What Factors Should You Consider?

To apply Figure 9.4 effectively, assess your situation across eight factors, grouped into three categories:

Control and ownership

These factors determine how much influence and differentiation you can retain.

- Customization: Do you need unique workflows, interfaces, or agent behaviors? Building allows full flexibility, buying limits options, and hybrid offers a middle ground.
- IP Ownership: Will owning the system's logic or training data provide lasting advantage? Building ensures full ownership; buying may limit it; hybrid can involve shared rights.
- System Control: How critical is control over how data is handled or how the system behaves? In-house systems offer maximum oversight; external platforms trade control for speed.

Ask: Do you need differentiation through unique features, tighter data control, or proprietary IP?

Feasibility and execution

These factors reflect the reality of getting your product out the door.

- Speed to Market: How quickly must you launch? Buying is fastest, hybrid is faster than building, but slower than buying.
- Cost: What's your available budget? Building typically requires more up-front investment; buying is often pay-as-you-go; hybrid can vary based on custom scope.
- Team Capability: Do you have the engineering, data science, or infra support to build and maintain the system? Internal builds require deeper technical muscle.

Consider: Are your timelines, budget, and staffing aligned with your chosen approach?

Risk & strategic alignment

These factors ensure your decision holds up over time.

- Risk and Security: How sensitive is your use case? Building limits vendor exposure; buying increases dependency risks; hybrid requires careful boundary-setting.
- Strategic Alignment: Does the sourcing strategy support your long-term product or platform vision? Build offers maximum extensibility; buy offers speed with constraints; hybrid allows evolution.

Reflect: Can you manage the risks—and does the approach support where your product is headed?

Tools

Build-Buy-Hybrid matrix

The *Build–Buy–Hybrid* Matrix, shown in Figure 9.5, helps you compare your sourcing options—building, buying, or creating a hybrid system—across the same three dimensions we explored earlier: Control and Ownership, Feasibility and Execution, and Risk and Strategic Alignment. It consolidates the eight decision factors into a simple, side-by-side comparison.

Dimension	Build	Buy	Hybrid
Control & Ownership	Full control over data, system behavior, and UX. IP stays in-house. Ideal for sensitive use cases or when differentiation is a priority.	Limited customization or control. IP and behavioral logic often reside with the vendor.	Moderate control—teams can wrap off-the-shelf systems with custom prompts, logic, or orchestration.
Feasibility & Execution	Slower time to market. Requires deep technical resources and budget.	Fastest to launch. Minimal setup or expertise required.	Medium pace. Leverages external capabilities, while tailoring where it matters.
Risk & Strategic Fit	Lower vendor risk. High scalability and alignment with long-term goals.	Higher risk of vendor lock-in or integration limits. Moderate scalability for broad use cases.	Third-party risk still present, but easier to adapt system over time as needs evolve.

Figure 9.5: Build-Buy–Hybrid matrix for GenAI system sourcing.

By using this matrix, you can quickly assess which path fits your product's constraints and ambitions. For instance, if you're working in a regulated industry with custom workflows and data sensitivity, building might be the right choice.

If you're validating a new use case or embedding AI in a non-core feature, buying a pre-built model or tool may suffice. And for many agentic systems—where you need to customize logic, tooling, or memory—a hybrid model offers flexibility without reinventing the wheel.

Choosing wisely at this stage helps avoid costly pivots later. You'll bypass the trap of overbuilding when a hybrid would do or the trap of underinvesting in a system that needs deeper control. This sourcing decision sets the stage for behavioral and UX design decisions to follow.

Consider Spotify. They've built their own core AI system—LLark—for deep music understanding and tailored playlist creation, while integrating third-party APIs for support features like voice-enabled DJ interactions in their AI DJ. This hybrid approach combines in-house models with API-based voice tech, enabling innovation without rebuilding every component. It also helps maintain data privacy and comply with regulations like GDPR—a smart balance between speed, control, and compliance (Jeong et al. 2023).

Decision 4: How Should Your GenAI System Behave?

With your GenAI system's architecture and sourcing strategy in place, the next critical decision is defining how your system should behave. This isn't just about what the system does but how it interacts with users, makes decisions, and manages autonomy, especially for agentic systems that operate independently. In my experience as a product manager, I've seen behavior design determine whether a GenAI product earns user trust or falls flat. I once advised a team that launched an agentic assistant with overly aggressive automation, scheduling meetings without user consent, which led to frustration and abandonment. That taught me that behavior is the linchpin of user adoption and system reliability. This section provides tools to help you shape your system's behavior, ensuring it's intuitive, trustworthy, and aligned with your strategic goals.

In Chapter 5, we explored interaction patterns like "chat," "copilot," and "agent," guiding you through execution-focused choices for user-facing designs. Here, we take a strategic perspective, focusing on the behavior of the systems—particularly agentic ones—before implementation begins. Next, we'll go over the key questions to ask when designing GenAI system behavior to ensure users embrace it rather than resist it.

Key Questions

How autonomous should your system be?

Autonomy determines how independently your system operates. Should it wait for user input, like a chatbot responding to queries, or act proactively, like an agent scheduling tasks? High autonomy, common in agentic systems, boosts efficiency but risks overstepping user expectations. Low autonomy ensures user control but may limit the system's value. Ask yourself: Does your use case require proactive action to deliver value, or is user oversight essential? For example, a customer support bot might need low autonomy to avoid errors, while a workflow assistant could benefit from proactive suggestions.

When should it escalate to a human?

No GenAI system is infallible, and defining when it should hand off a conversation or task to a human is crucial, especially in high-stakes contexts like legal or financial applications. Consider: What scenarios demand human judgment? How will the system recognize its limits? For instance, an agentic system might escalate if it detects ambiguous user intent or a sensitive task, preserving trust and safety. Clear escalation protocols ensure your system knows when to step back.

What risks must you mitigate?

GenAI behavior, particularly in agentic systems, introduces risks like hallucinations, unintended actions, or privacy violations. Ask: What could go wrong if the system acts incorrectly? How will you prevent harm? For example, an agent drafting emails might inadvertently share sensitive data if not properly constrained. Identifying risks up front allows you to design safeguards, such as approval prompts or restricted action scopes, to protect users and maintain reliability.

Tools

TAPE

The *TAPE* tool (task, autonomy, performance, escalation), shown in Figure 9.6, is a decision aid to use when designing your system's behavior. It helps you define the task scope, set autonomy levels, ensure performance meets user expectations, and establish escalation protocols, particularly for agentic systems.

Dimension	What to Ask	Why It Matters
Task	What specific tasks should the system perform?	Clear task boundaries prevent overreach and align with user needs.
Autonomy	How independently should the system act?	High autonomy boosts efficiency but risks errors; low autonomy ensures control.
Performance	What quality and speed standards must it meet?	Performance defines user satisfaction and system reliability.
Escalation	When and how should it hand off to a human?	Escalation protocols mitigate risks in ambiguous or high-stakes scenarios.

Figure 9.6: TAPE tool for designing GenAI system behavior.
© 2025 by the author of *Zero to GenAI Product Leader*.

By applying the TAPE tool you can craft behavior that balances autonomy, reliability, and trust, tailored to your system's objectives. You'll avoid designing systems that are either too autonomous or too rigid, a mistake I've seen undermine user trust and adoption. This decision ensures your system behaves in a way that users rely on, setting the stage for the UX and trust decisions we'll explore next.

When Google first introduced Duplex—a voice assistant that could autonomously call businesses to book appointments—it implemented high autonomy early on. Users weren't always aware they were speaking to AI, and Duplex occasionally misunderstood subtle conversational cues. These concerns led Google to revise the system. They began including voice disclosure (e.g., "I'm an automated assistant") and added escalation pathways to human operators for complex conversations. Duplex's evolution highlights why carefully designing autonomy and escalation frameworks is essential for GenAI systems (Leviathan and Matias 2018).

Decision 5: How Will Users Experience and Trust Your System?

With your GenAI system's architecture, sourcing, and behavior defined, the next critical decision is ensuring users have a seamless experience and trust the system. User experience and trust determine whether your system is adopted or abandoned, shaping its success in real-world use. As a product manager, I've learned that even the most advanced GenAI system can fail if users find it confusing, slow, or unreliable. This section provides tools to help you design an intuitive UX and build user trust, ensuring your system delivers value and inspires confidence.

In Chapter 5, we focused on tactical UX design, defining how your system behaves, communicates, and earns trust through clear interfaces, feedback loops, and safety mechanisms. Here in Chapter 9, we take a step back to ask the strategic questions that shape that experience: What kind of trust does your system need to earn? And what UX posture will support it from the start?

Key Questions

By the time you've mapped the system's architecture and behavior, it's tempting to jump straight into design. But before getting into interface details or microcopy, I find it helpful to pause and ask:

What kind of experience does this product need to succeed?

Different GenAI products call for different UX priorities. A code generation tool might need speed and easy editing. A healthcare assistant needs transparency and escalation. A travel bot might require error recovery and reassurance. There's no one-size-fits-all but there is usually a baseline your system has to meet to earn trust.

For instance, if your AI rewrites content, users should clearly see what's being edited and why. If your system summarizes conversations, users should be able to tweak or flag parts that feel off.

I usually focus on four:

- Latency – Where must the system feel instant? A design tool might need sub-second speed for a fluid experience.
- Interpretability – Where do users need to understand why it did what it did? If it's making decisions on their behalf, show your logic.
- Feedback – Where should they be able to correct or guide the output? Let them flag errors or edit results without having to start over.
- Trust – Where could the system lose confidence if it gets things wrong? Set clear boundaries, preview actions, and always give users control.

You won't optimize for everything but making intentional choices here means fewer surprises later.

Tools

The UX and Trust Design Matrix (Figure 9.7) is a decision tool that helps you translate UX and trust priorities into design choices, depending on whether you're building a tool, embedded agent, or copilot-style assistant.

System Type	UX Risk	Trust Requirements	Design Focus
Standalone tool (e.g., summarizer, document search)	Medium	Moderate – must be correct and editable	Speed and editability
Embedded agent (e.g., support assistant, planner)	High	High – must explain itself and escalate well	Transparency and fallback control
Copilot-style assistant (e.g., Microsoft 365 Copilot)	Variable	High – must align with user goals and context	Intent visibility and context-aware responses

Figure 9.7: UX and trust considerations by GenAI system type

This matrix isn't about rules, it's about setting the right expectations. A user-facing agent that takes initiative needs a different trust posture than a summarizer that quietly does its job in the background. The better you match UX priorities to system type, the more likely users are to feel confident, informed, and in control.

By clarifying what kind of experience your system needs to deliver—and which dimensions matter most—you'll build stronger foundations for trust and adoption before launch.

When Notion launched *Notion AI* in 2023, it didn't introduce a flashy chatbot or standalone assistant. Instead, it seamlessly embedded GenAI into users' existing writing workflows—a strategic move that emphasized clarity, speed, feedback, and trust. Users could highlight text and summon the AI to summarize, improve writing, or generate ideas, with each AI-driven action subtly labeled as an "AI suggestion." This transparency ensured users always understood when the system was assisting versus when they were editing directly, fostering confidence without confusion.

To minimize latency, Notion prioritized lightweight prompts and progressive output streaming, keeping AI responses almost instantaneous even for complex tasks. By showing partial outputs as they were generated, Notion reduced wait anxiety and maintained a fluid user experience. It also created simple feedback loops: users could immediately accept, edit, or discard AI outputs with a click, reinforcing a sense of control. If the AI's suggestion felt off, users could easily refine it or return to manual editing.

Crucially, Notion never over promised the capabilities of its AI. Rather than positioning it as a perfect assistant, the system framed itself as a copilot—helping users move faster while encouraging manual oversight. By prioritizing transparency, control, and reliability over automation hype, Notion built a GenAI experience users could trust (Notion 2023).

Decision 6: How Do You Scale and Monitor the System Over Time?

With your GenAI system's architecture, sourcing, behavior, and user experience defined, the next critical decision is how to scale and monitor it over time. Scaling ensures your system handles growing user demand and complexity, while monitoring maintains performance, cost-efficiency, and reliability. As a product manager, I've learned that proactive scaling and monitoring are essential to prevent a promising GenAI product from becoming a costly or unreliable burden. This section provides tools to help you scale responsibly and monitor effectively, ensuring your system thrives as it grows.

In Chapter 5, we explored scaling considerations like infrastructure and guardrails during system design execution. Here, we take a strategic perspective, focusing on planning and managing scaling and monitoring to sustain long-term success. These are the questions I ask myself to keep a GenAI system robust, and they'll guide you to build a product that grows without breaking.

Key Questions

How do you track versions and evaluate outcomes?

As your system scales, managing versions of models, data, and configurations ensures consistency and traceability. Ask: How will you track changes to your system? How will you evaluate its performance against user expectations? For example, a recommendation engine might need versioned models to maintain suggestion accuracy, with regular evaluations to measure user engagement. Clear versioning and evaluation processes prevent drift and ensure reliable outcomes.

How do you control cost, performance, and recovery?

Scaling often increases costs (e.g., compute, API calls) and strains performance, while failures can disrupt user trust. Consider: How will you balance cost and performance as usage grows? What recovery mechanisms will you implement for failures? For instance, an AI design tool might use cost caps on API usage and automated retries for failed tasks. Proactive cost management and recovery plans maintain efficiency and resilience.

What telemetry and optimization strategies will you use?

Telemetry—collecting data on system usage and performance—enables optimization to improve user experience and efficiency. Ask: What metrics will you monitor to understand system health? How will you optimize based on insights? For example, a brainstorming tool might track latency in content generation and optimize model inference to reduce delays. Robust telemetry and optimization ensure your system adapts to evolving needs.

Tools

VECTOR

The *VECTOR* tool (versioning, evaluation, cost, telemetry, optimization, recovery) helps you think holistically about what it takes to keep your GenAI system reliable as it scales. Shown in Figure 9.8, it frames the six core dimensions that often get overlooked once a system is in motion.

Dimension	What to Ask	Why It Matters
Versioning	How will you track model, data, and configuration changes?	Ensures consistency and traceability as the system evolves.
Evaluation	How will you measure performance against user expectations?	Maintains alignment with user needs and quality standards.
Cost	How will you manage costs as usage grows?	Prevents budget overruns while sustaining scalability.
Telemetry	What metrics will you monitor for system health?	Provides insights for performance and user experience improvements.
Optimization	How will you improve efficiency based on telemetry?	Enhances system performance and adaptability over time.
Recovery	What mechanisms will you implement for failures?	Ensures resilience and maintains user trust during disruptions.

Figure 9.8: VECTOR tool for scaling and monitoring GenAI systems.
© 2025 by the author of *Zero to GenAI Product Leader*.

It's not just about catching failures; it's about planning ahead. Are you versioning your models in a way that lets you roll back when needed? Do you have telemetry that goes beyond uptime to capture usage patterns and bottlenecks? Are you building in recovery options now instead of scrambling later?

VECTOR gives you a practical lens to stress-test your readiness across system health, cost efficiency, and long-term adaptability. I've seen teams avoid painful rework just by walking through this list early—before their GenAI system hits real scale.

By using VECTOR, you're not just shipping something that works today. You're building something that stays trustworthy, even as complexity grows.

That kind of long-term thinking is what helped GitHub scale Copilot from an early developer assistant into one of the most widely adopted GenAI tools in production today. Powered by OpenAI's Codex models, *GitHub Copilot* now delivers millions of coding suggestions daily across platforms like Visual Studio Code, GitHub Codespaces, and JetBrains IDEs. Scaling it successfully meant more than just infrastructure—it required disciplined attention to versioning, telemetry, and user feedback loops.

In a 2024 study, real-world projects saw significant efficiency gains: up to 50% time saved on documentation and autocompletion, and 30–40% savings on repetitive coding tasks, unit test generation, debugging, and pair programming. In public discussions, GitHub has shared that metrics such as suggestion acceptance rates, editing behaviors, and user engagement are critical for evaluating Copilot's ongoing quality and relevance. The system uses a scalable cloud backend tuned for low-latency delivery to ensure a smooth developer experience. GitHub has also added safeguards: detecting misuse, monitoring latency spikes, and managing load to maintain availability. In 2023, they opted into pilot telemetry programs letting users control data sharing to improve quality while preserving transparency and control (Huang et al. 2024).

Copilot's disciplined approach—tracking versions, tuning performance, and learning from user behavior—helped it scale without compromising trust or reliability. It's a strong example of VECTOR in action: scaling GenAI thoughtfully, without letting speed undermine confidence.

Conclusion

Building GenAI products is a series of critical decisions not just technical ones, but architectural, behavioral, operational, and strategic. In this chapter, we focused on how to reason through those decisions systematically, using simple tools to frame complex trade-offs without getting lost in technical detail.

The real challenge in GenAI product management is not knowing every framework—it is navigating uncertainty with structured judgment. Whether you are deciding where to invest, how your system should behave, or how to scale and monitor responsibly, decision fluency is the skill that sets enduring leaders apart.

By mastering structured decision-making, you will not only build more resilient GenAI products—you will lead them with greater clarity, adaptability, and long-term impact.

Chapter 10: Mapping Your GenAI Product Journey

"You cannot discover new oceans unless you have the courage to lose sight of the shore." — *André Gide*

In Chapter 4, we explored the skills, mindsets, and new ways of thinking that set GenAI Product Managers apart. Now, it's time to turn that foundation into action. This chapter is about you: mapping your unique strengths, finding your natural starting point in the GenAI product landscape, and planning your path forward. You already know what it takes to succeed, now let's figure out where to begin.

Breaking into GenAI product management isn't a straight path. It's more like navigating a dynamic, evolving landscape—one where your starting point, strengths, and aspirations all shape the route you take. Every GenAI product manager's journey is unique. Some of you may be approaching from a technical background, others from user research, consulting, business strategy, or early product roles. Where you begin matters, but it's not what defines you. What matters is how intentionally you map your growth—choosing which skills to build next, which areas to explore, and how to steadily expand your capabilities.

In this chapter, we'll map how your strengths align with real-world GenAI PM opportunities, how to choose the right starting point, and how to chart a path that evolves over time toward becoming a full-stack GenAI product manager. You don't need to have it all figured out today. You just need to know where you are and where you want to move next. Before you set sail on your own GenAI journey, it's worth pausing to hear a reflection from leaders who've already charted these waters.

__Amit Ghorawat, Director of PM at Reddit,__ confesses he once underestimated just how fast GenAI would evolve—and he still balances a "healthy paranoia" about the pace of change by diving into experiments rather than waiting for perfect clarity. This speed creates a fascinating paradox. On one hand, you realize you can never truly keep up with every single development. But on the flip side, that very same speed means it's never too late to dive in.

Amit also stresses his not-so-secret sauce: be super curious, share your early wins, and surround yourself with fellow tinkerers—a habit that will keep you experimenting when the pace of change feels overwhelming.

Keep these high-leverage insights in mind as you find your own "zone of strength" on the map and don't wait to learn them the hard way out in the wild.

Navigating Zones of Strength

Before we dive deeper, I want to give you a simple way to picture the landscape ahead. To help you find your starting point, I've created a visual map of the major zones of strength (See Figure 10.1) within GenAI product management.

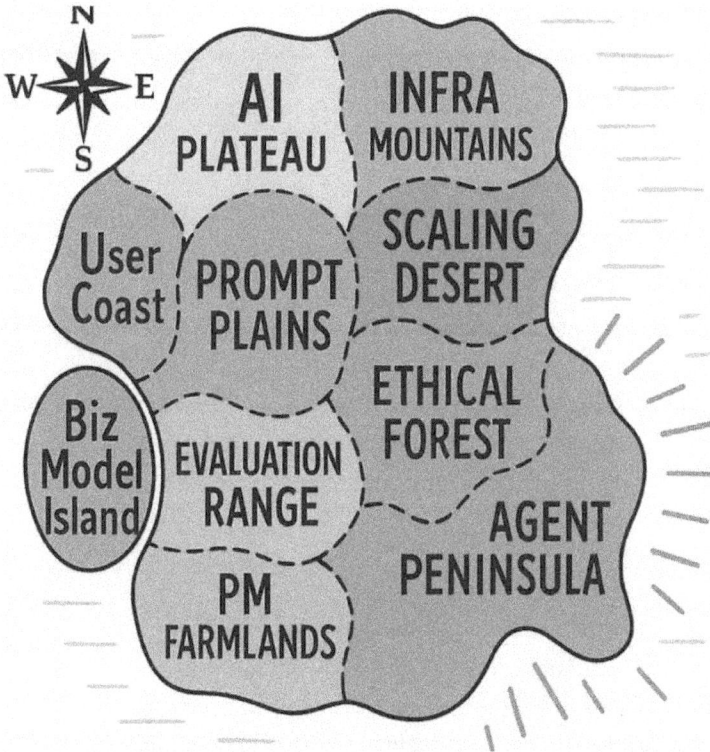

Figure 10.1: A simplified map of the major zones of strength within GenAI product management.
© 2025 by the author of *Zero to GenAI Product Leader*.

I didn't build this map to tell a story. I built it to give you a practical tool, something you can use to quickly see where your natural strengths already fit, and where you might want to stretch next.

I didn't just pick names like "Plateau", Mountains", "Plains", and "Desert" at random. They actually hint at what the work feels like in each area:

- *Plateaus are elevated, stable terrains—where PMs step back from daily sprints and steep climbs to focus on long-horizon model strategy, foundational research partnerships, and shaping core AI capabilities.*
- *Mountains are where deep technical complexity lives—the tough climbs, like scaling infrastructure or runtimes.*
- *Ranges are expansive terrains—where insights stretch across both time and scale. This is where GenAI PMs trace model behavior from first experiment to real-world performance, making continuous progress through data, feedback, and observation.*
- *Plains are all about fast moves—experimentation, iteration, and quick feedback loops happen here.*
- *Deserts signal scaling challenges—you'll need endurance and precision to grow products and systems that actually survive and thrive at scale.*
- *Forests are where careful, thoughtful decision-making matters most—like trust, fairness, and responsibility.*
- *Coasts, Islands, and Peninsulas point to specialized focus areas—like business models, ecosystems, or designing AI-powered experiences.*

You might notice that the Agent Peninsula has expanded its influence, reshaping neighboring regions like Ethical Forest and Scaling Desert. As AI agents mature, they're redefining how we think about infrastructure, ethics, scaling, and more. You don't need to memorize this map; it's here simply to give you a sense of the terrain and maybe help you feel a little less lost as you start charting your own path. Here's a brief orientation to each zone of strength.

AI Plateau

Plateaus are places for research and strategic planning for cutting-edge technologies such as new AI models, frameworks, and system prototypes. If your strengths lie in guiding early-stage research, coordinating prototype projects, and translating exploratory work into product strategies, the AI Plateau could be your zone. This isn't about incremental tweaks; it's where you:

- Work with researchers on proof-of-concept experiments for models, agentic frameworks, or tooling
- Validate emerging approaches and surface feasibility risks early
- Define multi-quarter development roadmaps that balance technical potential with business goals

You'll collaborate across research and engineering teams turning ideas (whether a novel architecture, an orchestration framework, or a data pipeline) into actionable plans. If you're energized by abstract technical challenges, willing to operate without perfect roadmaps, and motivated by long-term impact rather than immediate wins, AI Plateau could be your first stronghold in GenAI product management.

Typical roles in this zone include the following:

- Product Manager, Foundation Models
- Product Manager, AI Innovation & Prototyping
- Product Manager, AI Safety & Trust
- Product Manager, Research Tools & SDKs
- Product Manager, Emerging Model Architectures & Frameworks

Strength here means you help shape what "next" looks like scoping new capabilities, de-risking early research, and laying the groundwork for scalable AI products.

Infra Mountains

If you naturally lean toward systems thinking, backend engineering, or infrastructure optimization—and have a strong technical bent of mind—the Infra Mountains might be a great starting point for you.

This zone is all about building the foundations that make GenAI work at scale—not just tuning performance metrics, but designing the systems that let large models train, deploy, and serve users globally. It's for PMs who thrive at the intersection of performance, reliability, and cost—and who aren't afraid to get deep into cloud infrastructure, distributed systems, and MLOps. If you're the kind of person who enjoys solving hard technical problems behind the scenes— and influencing how engineering teams build for the future, the Infra Mountains could be your starting zone.

Here are some common role types where these skills are needed:

- Platform PM for AI services
- Model deployment and hosting PM
- AI infrastructure scaling PM
- Runtime systems PM for AI/ML workloads

Strength here means you'll likely own key platform capabilities, enabling scalable model training and inference, optimizing runtime systems, and ensuring that enterprise-grade AI products can perform under real-world load.

Prompt Plains

Plains are all about fast moves—experimentation, iteration, and quick feedback loops happen here. If you love rapid experimentation, creative problem-solving, and the idea of shaping AI behavior without retraining models, the Prompt Plains might feel like home. This zone is all about crafting, engineering, and refining prompts (and retrieval-augmented generation systems) to guide how GenAI models behave in the real world. It's for PMs who are part artist, part scientist—tuning prompts, designing interaction patterns, and helping models respond more usefully and safely to user needs.

Typical roles in this zone include:

- Product Manager, Prompt Engineering
- Product Manager, Retrieval-Augmented Generation (RAG)

Strength here means you'll build prompt libraries, design RAG pipelines, and iterate on system behaviors through prompt experiments—powering immediate impact without a full retrain.

Scaling Desert

Deserts signal scaling challenges—you'll need endurance and precision to grow products and systems that actually survive and thrive at scale.

If you're driven by operational rigor, data-driven optimization, and the strategic orchestration of systems under stress, the Scaling Desert might be your domain. This zone focuses on designing and iterating the processes, pipelines, and performance guardrails that allow GenAI solutions to handle real-world variability and volume. It's for PMs who excel at analyzing usage patterns, reducing latency, and optimizing cost, ensuring products stay resilient as adoption grows.

Typical roles in this zone include:

- Product Manager, MLOps & Workflow Automation
- Product Manager, AI Scalability & Performance
- Product Manager, Reliability & Monitoring

- Product Manager, Cost & Efficiency Optimization

Strength here means you'll own the systems-level playbook for scale—tuning performance, reinforcing reliability, and optimizing resources so GenAI products can endure and adapt under heavy load.

Sunny Tahilramani (Gen AI Product Leader, Google) underscores why this matters: mastering unit economics—the cost of inference, latency, and model-stack complexity—is a fundamental product decision, not just an engineering detail. He also wishes he'd known early that the best GenAI products don't stop at features—they evolve the problem itself. Designing modular, extensible systems lets users push the boundaries as model capabilities grow.

Ethical Forest

The Ethical Forest is a zone of strength for those who believe that how we build matters just as much as what we build.

If you feel a natural pull toward questions of fairness, transparency, user safety, and trust, you're already wired for this zone. Your instincts aren't just nice-to-haves; they're becoming critical muscles in the next generation of AI leadership.

Operating in the Ethical Forest means more than setting guidelines or checking compliance boxes. It's about weaving responsible practices directly into how GenAI products behave, learn, and evolve. You're not standing on the sidelines, you're shaping the rules of engagement for a world where AI touches nearly every part of life.

You'll take on challenges like mitigating bias in model behavior, designing systems that can explain their decisions, safeguarding user trust under uncertainty, and building guardrails that evolve as AI becomes more autonomous. If you want to be part of building AI that people can actually trust—at a time when the stakes are only getting higher, the Ethical Forest could be your starting ground.

Here are some common role types where this zone of strength shines:

- Product Manager, Responsible AI
- Product Manager, AI Safety & Governance
- Product Manager, Fairness & Explainability

Organizations across every sector from tech giants to healthcare to finance are realizing that scaling GenAI without scaling responsibility is a fast track to failure.

That's why PMs who lead from the Ethical Forest aren't just seen as risk managers, they're increasingly viewed as essential architects of sustainable innovation.

Finding your footing here doesn't mean giving up speed or ambition. It means building the kind of future that's worth racing toward.

User Coast

The User Coast is where product instincts meet human intuition.

If you naturally think about how people *feel* when they interact with technology, if you're drawn to designing experiences that are seamless, empowering, and trustworthy—this could be your first zone of strength.

Operating here means crafting how users discover, interact with, and ultimately trust AI systems. You're not building models behind the scenes, you're shaping the front door, the handshake, the conversation.

You define how intelligence feels when it reaches the user. You'll take on challenges like designing intuitive prompts and interfaces, embedding transparency into the user journey, enabling feedback loops, personalizing interactions, and ensuring that every AI touchpoint builds (rather than erodes) trust.

If you have strong UX instincts, a bias toward user-centered thinking, and a passion for making AI accessible and human, the User Coast could be where your journey begins.

Common role types where this zone of strength shines include the following:

- UX for AI product manager (conversational design, human–AI interaction)
- Trust and safety PM (interaction layer)
- Personalization and user behavior PM

In today's GenAI world, how users experience AI is as critical as what the AI can do. Companies know this—whether it's building chat interfaces for enterprise co-pilots, AI companions for creators, or dynamic help systems for apps. That's why PMs on the User Coast aren't just polishing the surface, they're building the bridges that turn raw intelligence into everyday value, earning and maintaining user trust through transparency.

Anchoring yourself here means not just building better features, it means building better relationships between humans and AI.

User Trust in Practice

As **Saty Das, Staff PM at TikTok Shop,** learned early on, "One of the best decisions I made was to stay close to design and customer success teams. Even the best product ideas fail if users don't trust them or understand them." **Rocky Zhang, Gen AI Product Leader at Airbnb,** discovered that explainability isn't just an engineering checkbox—it's a core product design challenge. Surface the "why" from day one so users understand and trust your AI.

PM Farmlands

The PM Farmlands isn't just one specialization, it's the source that every GenAI PM relies on.

If you already have strong instincts around goal-setting, prioritization, customer discovery, and iterative development, you're in a great position. It means you already know how to turn uncertainty into action—even when the path isn't perfectly clear.

Strength in the PM Farmlands gives you choices: you could build a career as a generalist GenAI PM, leading end-to-end AI products without needing deep specialization. Or you could branch into another zone of strength like diving deeper into technical systems (Infra Mountains), user interactions (User Coast), ethical design (Ethical Forest), or agentic product orchestration (Agent Peninsula). Where you go next depends on your curiosity and ambition, but a strong foundation in product management will support you no matter which path you choose.

Common role types that rely heavily on this foundation include the following:

- Product Manager, GenAI Core Products
- Product Manager, Growth & Adoption (GenAI)
- Product Manager, [Industry] AI (e.g., Healthcare AI, Finance AI)

Having strong roots in the PM Farmlands means you already know how to cultivate products thoughtfully and now, you're ready to explore the wilder landscapes that GenAI is opening up.

Biz Model Island

Biz Model Island is where strategic instincts meet the GenAI economy.

If you're energized by thinking about how products actually create value, how they get priced, packaged, launched, and scaled, this could be your first zone of strength. It's not just about building the technology; it's about making sure it has a sustainable, thriving place in the world.

Strength in Biz Model Island means you're tuned into the realities of business models, monetization strategies, and go-to-market dynamics. You naturally think about questions like: *Who pays for this? What's the right pricing model? How do we scale distribution?*

In the GenAI world, these aren't afterthoughts, they are part of the product itself. If you enjoy shaping pricing strategies, experimenting with value metrics, designing revenue streams, or building ecosystem partnerships that accelerate adoption, Biz Model Island might be your natural home base.

Here are some common role types where this zone of strength shines:

- Product Manager, Monetization & Pricing Strategy
- Product Manager, Go-to-Market Strategy (GenAI)
- Product Manager, Ecosystem Growth & Partnerships
- Product Manager, Commercialization Strategy (AI Startups)

Working from Biz Model Island doesn't mean you'll only think about dollars and deals—it means you'll help craft products that are not just innovative, but sustainable. In a GenAI world full of technical breakthroughs, your strength is ensuring those breakthroughs turn into real-world success stories.

You're not just helping build AI products. You're helping them survive and thrive out in the wild.

Evaluation Range

Ranges are expansive terrains where insights stretch across both time and system layers.

This zone of strength spans offline evaluation, live observability, data curation, and feedback loops. You see AI not as something we ship and forget, but as something we grow—through careful measurement, user alignment, and iteration.

If you're passionate about understanding how AI systems—be they foundation models, retrieval pipelines, or multi-step agents—perform in real conditions, the Evaluation Range is your domain.

You'll build robust evaluation frameworks and feedback loops that carry systems from prototype through production, ensuring AI delivers consistent quality, adapts to new data, and stays aligned with business and user goals. You'll:

- Define end-to-end evaluation frameworks for models, agents, and AI applications
- Design and operate benchmarking pipelines: offline tests, simulation environments, A/B experiments
- Build observability and telemetry systems to monitor performance, bias, drift, and user satisfaction
- Close loops between data insights and product roadmaps to drive continuous improvement

Typical roles in this zone include the following:

- Product Manager, System Evaluation & Benchmarking
- Product Manager, AI Monitoring & Observability
- Product Manager, Performance Analytics & Metrics
- Product Manager, Telemetry & Data Pipelines

Strength here means you're the "eyes on the range," making sure every AI capability—no matter how complex—stays reliable, fair, and effective as it scales and evolves.

Agent Peninsula

Peninsulas are specialized outposts where agent orchestration and integration frameworks take shape.

As agentic capabilities spread into every zone—from evaluation (self-testing agents) and user experiences (conversational flows) to infrastructure (agents provisioning compute) and monetization (agents-as-a-service)—the Agent Peninsula remains the dedicated outpost for anyone whose core focus is building the platforms and tools that make autonomous, multi-step workflows reliable.

If you're drawn to designing how agents plan, coordinate, and act on users' behalf, this is your zone. You'll:

- Design APIs and SDKs that let developers assemble, extend, and customize agent behaviors
- Specify orchestration layers for task coordination, agent-to-agent handoffs, and error recovery
- Establish governance, safety guardrails, and monitoring models tailored to autonomous workflows
- Integrate agent frameworks with infrastructure, data pipelines, evaluation systems, and user-facing interfaces

Typical roles in this zone include the following:

- Product Manager, Agent Orchestration Platforms
- Product Manager, Autonomous Workflow Systems
- Product Manager, Multi-Agent Frameworks
- Product Manager, Developer Platforms for AI Agents

Strength here means you own the "glue" between services, turning individual agent components into cohesive, goal-driven systems and ensuring those systems perform predictably wherever they touch the rest of the AI landscape.

Finding Your Zone: Build from Strength and Curiosity

Each role you pursue will draw from different combinations of these clusters. Take a moment to glance at Figure 10.1 and ask yourself:

- *Where am I strongest today?*
- *Which areas could I grow into next to expand my skills as a GenAI product manager?*
- *Or, if I'm just beginning, which region naturally fits the strengths I already have?*

For example, if you come from a data science background, you might find yourself strong in AI Foundations and Infra Mountains, but less experienced with user experience or business models. If you're a UX designer, your strengths might lie along the User Coast and PM Farmlands, while technical foundations could be areas for growth. Business consultants often start out strong on Biz Model Island, but need to build confidence across AI Foundations and Prompt Plains. Traditional PMs might find themselves comfortably anchored in PM Farmlands, while needing to grow technical and ethical expertise. And infrastructure engineers typically stand tall on Infra Mountains and Scaling Desert, but may need to build user-centric and strategic skills.

The point isn't to judge where you are, it's to locate yourself honestly.

- If you already have some exposure to GenAI systems, you can use the map to identify adjacent regions where you could deepen your skills.
- If you're transitioning into GenAI product management for the first time, you can use the map to find your most natural starting point, the region that best aligns with your existing strengths.

Either way, the goal is the same: start intentionally and build momentum one step at a time. But skills and experiences are only half the story. As you reflect on your starting point, don't just think in terms of titles or technical knowledge. Think about two things:

- The strengths you've already built
- The kinds of problems that genuinely energize you

Start with where you already have momentum. Maybe you haven't worked on GenAI products yet—but perhaps you've led complex infrastructure projects, shaped user experiences for emerging technologies, or navigated high-stakes product launches. These experiences are not side notes. They're real strengths that can translate into your next move.

When in doubt, ask yourself: *"Which regions of the map could benefit from the instincts and skills I already have?*

At the same time, remember: Your career isn't just built on skills, it's built on energy. What kinds of problems would you stay up late reading about, even if nobody asked you to? If system architecture excites you, Infra Mountains or Scaling Desert may feel like a natural fit. If building intuitive, trustworthy technology for humans sparks your imagination, the User Coast might be calling you.

Strength + genuine curiosity = a powerful starting point.

You don't have to be a master yet. You just need a direction where your existing momentum and your growing passion can start to align. Reflect a bit before moving forward: What past experiences—and what types of problems—make you feel most energized to grow as a GenAI product manager?

Amit Ghorawat (Director of PM, Reddit)'s Pivotal Lesson: "Don't shy away from impactful problems, even if you lack initial domain expertise. Being an 'outsider' can become your superpower—first-principles thinking and rapid learning let you question norms and drive innovation."

Mini-Case: Amit Ghorawat's Outsider Advantage

Amit kicked off his career in US healthcare—despite zero domain expertise—by reverse-engineering jargon, information flows, and financial levers. That "outsider" lens became his superpower: he dissected complex systems at Harvard Business School, then co-founded Bicycle Health to tackle the opioid crisis, before leaping into advertising and mastering multi-sided business models. Today, he applies that same first-principles, rapid-learning approach to GenAI: dive in, question assumptions, and let fresh perspectives uncover the biggest impact.

You don't have to be a master yet. You just need a direction where your existing momentum and your growing passion can start to align. Reflect a bit before moving forward: What past experiences—and what types of problems—make you feel most energized to grow as a GenAI product manager?

Start Smart: What Matters Beyond Strength

Finding your starting zone of strength is important but how you pick your first real-world role matters just as much. It's not just about where you can show early strengths. It's about giving yourself the best opportunity to learn deeply, build momentum, and grow your impact over time.

Before you jump into a specific opportunity, it's worth pausing to think bigger: Which early choices will set you up not just to land a role—but to thrive, evolve, and shape the kind of GenAI PM career you actually want?

Here are a few key principles to keep in mind as you choose your first GenAI PM role:

Check Your Motivation & Master the Basics

Before anything else, make sure your "why" and your foundation are rock solid. As **Mark Cramer (Sr PM at Meta)** advises: *"If you want to set yourself up for long-term success, double-check your motivation and then master the basics."*

Team Quality over Company Brand

Early on, *who* you work with matters even more than *where* you work. Look for teams where you can learn from strong PMs, engineers, and AI researchers—even if the company isn't a household name yet. A great team will accelerate your growth far faster than a flashy logo ever could.

Learning Surface Area

Choose environments that expand your real-world intuition about GenAI—where you can encounter live model behavior, user interactions, ethical trade-offs, and scaling hurdles. The wider the surface area you operate in early, the faster GenAI complexity will start to feel second nature.

Strategic Compounding

Think about how the instincts you build now can stack and expand over time. A smart starting point doesn't box you in, it opens new frontiers you can grow into.

For example, if you start in Prompt Plains, you might later deepen your impact across Evaluation Range or Infra Mountains. If you begin in PM Farmlands, you could expand into Ethical Forest or Biz Model Island over time.

Always ask yourself: *"Does this move expand my surface area for future zones too—not just today, but a year or two from now?"* Careers aren't static. They're living ecosystems you cultivate through thoughtful moves.

Bias Toward Action—Not Just Research

It's easy to get stuck in teams that endlessly analyze GenAI without shipping much. Prioritize environments where you'll actually launch features, experiment with users, and move models into production.

Building intuition about what survives contact with real users is priceless—and irreplaceable.

Tolerance for Ambiguity and Change

GenAI roadmaps can shift fast. Models improve. Regulatory winds change. Choose a space—and a mindset—where you're excited, not paralyzed, by change. The best GenAI PMs treat uncertainty as part of the adventure, not an obstacle.

Shambhavi Rao (Gen AI Product Lead at Google Search)'s "No One Has It All Figured Out": *"No one has it all figured out and that's okay." Embrace early ambiguity by making a credible first move rather than waiting for perfect clarity.*

Mission Alignment Matters

Finally, remember: GenAI is powerful—but where you apply it matters. Whether it's education, healthcare, creativity, enterprise tools, or AI safety, pick a mission you genuinely care about. It's much easier to navigate the ups and downs when you're building something you actually believe should exist in the world.

Selena Zhang (AI Product Leader at a global social media company)'s Dual-Lens Approach: *"Develop both technical literacy and domain fluency: understand how models work, but just as importantly, know when and how to set boundaries. Build with curiosity—but always anchor features in real-world safety and trust constraints."*

Feasibility and Credible First Moves

Ambition matters but realism wins early momentum. Ask yourself: *"What is the most credible first step from where I am today?"* You don't have to jump straight into building autonomous multi-agent systems. A smart, reachable first move will create success stories that open bigger doors later.

Recap: Your GenAI PM Career Is a Journey

Choosing your first GenAI PM role isn't just about where you're strong today. It's about where your instincts, your curiosity, and your motivation align with the challenges ahead—even if you're still early in your journey.

Strength gives you a starting point. Curiosity and momentum carry you forward.

Choosing your first region isn't about chasing the trendiest space. It's about matching your starting point with real-world opportunity, while keeping future doors open.

You don't have to get it perfect. You just have to make an intentional, thoughtful move—one that sets a strong foundation for growth.

Now that you have a clear idea of how to choose your starting point wisely, let's bring it to life. In the next section, I'll walk you through a few sample GenAI PM journeys, so you can see how different starting points and growth paths come together in the real world.

Real-world GenAI PM Examples

You've seen how to find your starting point. You've seen how to choose your first role thoughtfully.

Now, let's bring it to life with real-world examples. These sample journeys aren't meant to be rigid templates. They're sketches—showing how different starting strengths, curiosities, and first moves can evolve into broader GenAI PM careers over time.

As you read them, don't just look for a "perfect match" to your own background. Instead, notice the patterns:

- How people build momentum from where they already are
- How adjacent growth opens up new opportunities
- How intentional first steps lead to unexpected (and exciting) destinations

Use these journeys as inspiration; don't feel pressure to imitate them. Your path will be your own. But seeing how others navigate the map can help you chart a smarter course for yourself.

Real-world Journey 1: From Field Specialist to GenAI Infra PM

(Seokjin Han, Senior PM @Microsoft)

Zone of strength

Seokjin spent nearly seven years as a Global Black Belt AI Specialist at Microsoft—a deeply technical, field-facing role. His natural strength wasn't traditional product management; it was hands-on experience with deploying, scaling, and troubleshooting real-world AI systems.

He already understood, from the trenches, how models broke in production, how inference costs exploded, how reliability risks showed up in live customer environments.

That foundation positioned him closest to the Infra Mountains and Scaling Desert—the zones where AI systems must survive the messy reality of the world.

Starting smart

When Seokjin transitioned into product management, he chose a role that built on his field strengths rather than abandoning them. He stepped into a PM role inside the Azure AI Platform team—focusing on model inferencing, deployment, and runtime optimization.

It wasn't just a fit for his skills; it was a fit for his lived intuition about customer needs, technical bottlenecks, and real-world operational challenges.

Instead of chasing trendy GenAI features or brand-new UI products, he started smart—he picked a space where he could contribute from day one *and* expand rapidly.

What matters beyond strength

His early environment played a critical role in amplifying his growth. By working closely with some of the strongest engineers building Azure's core AI runtime infrastructure, he quickly deepened his technical expertise—not just in isolated areas, but across the broader system.

His day-to-day work exposed him to the full learning surface of production AI: scaling challenges, cost optimization, runtime performance, and the evolving standards for inferencing at cloud scale.

Over time, what began as field-tested instincts about deployment matured into strategic leadership—helping shape not just individual model launches, but the way Azure itself scaled AI workloads for customers around the world.

Seokjin's journey is a reminder: *Leveraging your real-world strengths—and choosing an environment that stretches you thoughtfully—can create a compounding advantage in GenAI product management.*

Real-world Journey 2: From UX Design to AI-Driven Product Strategy

(Satyajit Das, Staff Product Manager @TikTok Shop)

Zone of strength: Design intuition meets system thinking

Satyajit's journey into product management wasn't a leap, it was a thoughtful expansion. With over a decade of experience in UX and interaction design across SAP, Infosys, and IBM, he developed a deep intuition for how users navigate complexity. His early strength wasn't writing PRDs or defining revenue models, it was crafting systems where users felt confident, informed, and in control.

That background gave him a unique edge when he transitioned into product roles: he didn't just define features; he saw the whole experience. He understood how product decisions ripple through workflows, support systems, and trust signals. His strength naturally mapped to User Coast and Prompt Plains where usability, comprehension, and AI interpretation shape product success. But he didn't stop there.

Starting smart: Designing systems, not just screens

As Satyajit moved into PM roles at SAP Ariba and then Meta, he applied his design-driven mindset to high-stakes enterprise and B2C systems. At Meta, he led product efforts on conversational platforms enabling businesses to create contextual, AI-supported chat experiences. He managed cross-platform SDKs, scaled proactive engagement features, and worked with researchers to drive intelligent automation in customer support.

This wasn't just UI work anymore. He was operating at the intersection of design, AI, and infrastructure—managing APIs, data pipelines, and product roadmaps that served millions. These experiences positioned him closer to Agent Peninsula and Scaling Desert, while still staying grounded in the usability-first lens that defined his early career.

What matters beyond strength: Stretching into GenAI product leadership

Now at TikTok Shop, Satyajit leads seller communication systems that serve 400K+ merchants across global markets. He's delivered intelligent automation tools like intent-based bots and agent-assist systems—bridging the gap between human support and AI-driven resolution. While not every solution is GenAI to-day, the architecture, decision-making, and design considerations closely mirror the complexity of modern GenAI deployments.

What sets his journey apart is his commitment to expanding his PM depth without losing his design roots. Rather than treating AI as a distant technical layer, he's embedded it into workflows that drive real business value—improving CSAT, resolution rates, and seller responsiveness across regions.

Now, as he deepens his GenAI expertise, he's building from a powerful founda-tion: a PM who doesn't just ask *"What can this model do?"*, but also *"How will users trust it, adopt it, and succeed with it?"*

Evolving the zone: From user coast to agent peninsula

Satyajit's path is a model of layered growth—starting in the creative space of UX, expanding through infrastructure, then stepping into the orchestration challenges of intelligent platforms. His progression didn't involve abandoning his early strengths, it involved recasting them as strategic superpowers in a new domain.

In a world where GenAI products increasingly rely on human-AI interaction, his ability to translate behavior into design—and design into trust—is more valuable than ever. His growth arc shows how designers can become platform PMs, and how systems-thinking can be a launchpad into GenAI product leadership.

The bigger lesson: Don't skip steps — build layers

Satyajit's story isn't just about a title change. It's about compounding capability. He didn't try to become a GenAI PM overnight. He built domain fluency in de-sign, technical systems, APIs, and AI-driven automation—and continues to grow outward from there.

For anyone moving from UX, research, or experience design into GenAI product roles, his path is proof that you don't need to start from scratch—you need to start from where you're already strong and build deliberately outward.

Real-world Journey 3: Start with a Wedge, Expand into Adjacent Area

(PM X, GenAI Product Manager @Major cloud AI provider)

When you look at career moves from the outside, they can seem perfectly linear. But real growth usually happens by starting with a wedge—one area where you have momentum—and then expanding deliberately into adjacent zones. PM X's journey into GenAI product management offers a clear, real-world example of how this strategy plays out over time.

Start with a wedge: zone of strength

When PM X transitioned into the Cloud AI Platform team, they didn't spread their efforts across every part of the ecosystem. Instead, they started with a clear, powerful wedge: leading monetization strategy and go-to-market efforts for GenAI models.

Drawing from their strong foundation in AI technical sales and product marketing, PM X played a key role in shaping the models-as-a-service monetization strategy and guiding the global launch of high-profile models from third party model developers and model aggregators.

This focus let them anchor in their area of expertise—at the intersection of business strategy, AI commercialization, and cross-functional execution—rather than chasing every new trend.

Expand your surface area

As PM X embedded deeper into the cloud AI platform team, they didn't stay confined to their original wedge.

Their environment exposed them to a broader set of challenges—not just monetization, but developer onboarding, model lifecycle management, infrastructure scaling, and evolving inference standards.

They then took on ownership of the product strategy for a developer sandbox environment—designing experiences that let engineers prototype agent workflows and multimodal pipelines. This move shifted their focus outward: from pure business-model strategy into the heart of developer experience design, platform evolution, and fast-cycle experimentation.

By growing their surface area beyond that initial strength, PM X positioned themselves to influence not only how GenAI offerings were monetized, but also how they were discovered, tested, and brought to life by users.

Grow through adjacent zones

PM X's progression wasn't a random expansion; it was an intentional journey through adjacent zones of growth. Starting from Biz Model Island, they moved into Prompt Plains and then into User Coast.

At every step, they didn't abandon their original expertise; they layered new dimensions on top of it. They brought business acumen into developer-focused spaces and go-to-market instincts into infrastructure conversations.

Each move let them stitch together a broader, more durable skill set—connecting commercialization, user interaction, and production-scale platform realities. Rather than leaping into a completely foreign zone all at once, PM X expanded naturally outward—building credibility, trust, and expertise as they went.

Compounding momentum over time

PM X's journey stands out for how deliberately they compounded momentum. Rather than treating their early monetization work as a silo, they used it as a launchpad to expand into more complex, strategic roles across the GenAI value chain.

By starting with demonstrated strengths, choosing environments that broadened their exposure, moving into adjacent zones, and continuously layering new experiences, PM X turned an initial wedge into a wide sphere of influence.

Today, they help shape not only which GenAI models the platform offers, but also how those models are discovered, adopted, and scaled by developers worldwide.

Their path shows that early strength isn't about finding a "perfect" starting point—it's about choosing wisely and growing outward with intention, adaptability, and curiosity. In a fast-moving AI landscape, that mindset is the blueprint for long-term success.

Real-world Journey 4: From Space Scientist to GenAI Infra PM

(Kriti Faujdar, Senior Product Manager @Microsoft)

A career rooted in rigor

Before stepping into the world of GenAI, Kriti spent over seven years as a scientist at the Indian Space Research Organization (ISRO) where she was part of the mission-critical team responsible for launching and operating satellites. Her early career wasn't just technical, it was mission-critical. Precision, stability, and scalability were a necessity.

These experiences shaped her systems mindset—thinking in frameworks, planning for edge cases, and building for long-term resilience. That mindset has carried through every product role since.

Turning technical depth into a springboard

After completing her MS in Computer Science in the U.S., Kriti joined Microsoft as a PM intern, then transitioned into a full-time role at Microsoft Security Response Center. There, she worked on Azure's vulnerability management and bug bounty programs—balancing security, scale, and operational speed.

As a side project, she collaborated with engineers to build a deep learning model for log anomaly detection—applying techniques from her MS coursework to solve real-world challenges. These projects gave her a strong foundation in building products that operate behind the scenes—quietly, reliably, and at scale. While the domain was security, the skills she built became a springboard into AI infrastructure product management.

Lay the groundwork, then push boundaries

Today, Kriti owns the backbone of GenAI at Azure AI—onboarding, deployment infrastructure, and developer tooling that make foundation models (from Mistral, Meta, Nixtla, and others) accessible, scalable, and production-ready.

She built this expertise step by step—first at ISRO, then in security and incident response, and through curiosity-driven side projects in emerging AI spaces. Along the way, she embedded responsible deployment and real-world usability into every layer of the stack.

Kriti's journey reminds us:

By solving hard, foundational problems first and then deliberately expanding into adjacent areas, you can turn deep technical expertise into a durable, infrastructure-first path in GenAI product leadership.

Wondering how your own background might translate into GenAI product roles? Check out Appendix D: Transition Paths to GenAI PM by Background—it breaks down tailored entry paths for developers, designers, marketers, students, and more.

Common Challenges on the Path to GenAI Product Management

Let's be real, this transition isn't always smooth. Even with clear strengths and the right mindset, you're likely to hit a few bumps. Not because you're doing it wrong, but because breaking into something new—especially something as dynamic and fast-moving as GenAI—comes with friction.

Here are a few common struggles I've seen (and personally experienced).

The "You Don't Have GenAI Experience" Loop

You need experience to get the job. But you need the job to get experience. Classic catch-22. The way out? Show GenAI *thinking*, not just credentials. Highlight how you've worked with ambiguity, shipped complex products, or explored AI in side projects, mockups, writing, or cross-functional work. Don't just tell people you're interested. Show them you're already learning out loud.

Vague Job Descriptions

Many GenAI PM roles are still being defined, and in some cases, the hiring team doesn't even know exactly what they need. Your job? Help reduce that ambiguity. Frame yourself clearly. Map your strengths to what the team is likely struggling with—and make your curiosity a feature, not a liability.

Skill Gaps That Feel Daunting

It's easy to feel like you need to become an AI researcher overnight. You don't. You just need to build enough fluency to ask good questions, partner with engineers, and make product calls with confidence. Focus on the highest-leverage concepts: model behavior, prompting, evaluation, deployment, feedback loops.

Signal vs. Noise on Your Resume

Lots of applicants list "prompt engineering" or "took an AI course." That's not a story, it's a checkbox. Hiring managers are looking for signals that you know how to build, how to learn, and how to drive outcomes. A well-framed case study, a product teardown, or a shipped experiment says more than ten certificates ever could.

Confidence Dips

Imposter syndrome tends to spike just as you're leveling up.

That feeling of "I'm not ready" everyone has it. What matters more is forward motion. Keep building, keep shipping, keep asking better questions. You're not trying to prove you're perfect. You're trying to earn the next conversation.

Every successful GenAI PM I've talked to, even the ones now leading agent teams or running AI platforms—started with uncertainty. What set them apart wasn't perfection. It was persistence, direction, and the willingness to learn out loud.

Expanding Beyond Your First Region: Becoming a Full-Stack GenAI PM

By now, you've seen how finding your starting point matters—and how real-world journeys often begin with a wedge: a zone of strength where you already have momentum.

But here's the bigger truth: Your first region is just that—a first region. The best GenAI product managers don't stay anchored in just one zone forever. They expand thoughtfully. They grow in range. They become what I call full-stack GenAI PMs—people who can move fluidly across technical, user, orchestration, and business layers of the GenAI landscape. Let's talk about how that happens.

Your First Wedge Isn't Your Last Identity

If you look back at the journeys we just explored, you'll notice a pattern.

Seokjin Han started by translating deep technical field insights into platform PM leadership, moving from deployment experience to owning runtime evolution. Thasmika Gokal began with monetization and GTM strategy—but didn't stop there. She expanded into developer experience design, model experimentation, and broader platform growth.

They didn't abandon their starting strengths. They *built on them*—step by step—expanding into adjacent zones that made their impact broader and more durable. The same will be true for you. Your starting zone—whether it's AI Foundations, Infra Mountains, User Coast, or Biz Model Island—is your foundation. It gives you a platform to stand on. But over time, you'll want to reach beyond it.

What a Full-Stack GenAI PM Looks Like

You don't have to become an expert in everything overnight. But the GenAI PMs who thrive long-term tend to develop layered fluency across multiple parts of the map.

A full-stack GenAI PM might do any of the following things:

- Understand how AI models actually work, including their strengths and limitations (AI Plateau).
- Design intuitive, trustworthy, human-centered experiences (User Coast).
- Navigate infra scaling, cost trade-offs, and production deployment (Infra Mountains and Scaling Desert).
- Shape monetization strategies and ecosystem growth (Biz Model Island).
- Build and orchestrate agentic workflows (Agent Peninsula).

No one starts with all of this. But each thoughtful move expands your surface area—and your ability to shape meaningful AI products at scale.

How to Expand Thoughtfully

You don't need to chase every new buzzword or tech trend. Instead, expand adjacently and deliberately:

- Grow sideways: If you started in Prompt Plains (prompt behavior shaping), you might next deepen into Evaluation Range (model behavior evaluation) or Ethical Forest (responsible AI considerations).
- Stack strategically: If you anchored in the Infra Mountains, you might next pick up orchestration skills from Scaling Desert or agent design from Agent Peninsula.

- Follow curiosity, not panic: Let your natural interests pull you into new regions. When you're genuinely excited about the work, learning compounds faster—and feels less like a chore.

Think in seasons. Every 6–12 months, ask yourself: *"Where can I add one more dimension to my capabilities?"* Not by abandoning your wedge but by thoughtfully adding to it.

What Makes Expansion Hard—and How to Handle It

Becoming a full-stack GenAI PM isn't just about ambition. It takes stamina, adaptability, and honesty about the friction you'll face along the way. Here are some of the common hurdles you'll encounter—and how to grow through them.

Switching from Builder to Strategist

If you come from engineering or data science, expanding into user or business zones might feel soft or vague. The work is less about correctness—and more about resonance, trust, and behavior.

What helps: Co-own a user-facing feature. Shadow UX researchers. Sit in on roadmap strategy meetings even as a listener.

Navigating Technical Complexity Without Deep Fluency

If you started on the User Coast or Biz Model Island, stepping into Infra Mountains or Evaluation Range can be intimidating.

What helps: Pair deeply with engineers. Ask "why" until you find patterns. Frame technical conversations around product risks and opportunities—not trivia.

Balancing Systems Thinking with Human Thinking

The deeper you go into infrastructure or orchestration, the easier it is to lose sight of the person on the other side.

What helps: Schedule regular user testing. Shadow customer support. Rotate between infra-heavy and user-facing workstreams.

Letting Go of Control as Products Get Smarter

As you start working on adaptive systems, outcomes become less deterministic. This can frustrate PMs who are used to tightly specifying behavior.

What helps: Reframe your role. You're not specifying outcomes, you're designing systems with boundaries, feedback loops, and incentives.

Internal Labels That Lock You In

Sometimes, your team sees you as "the infra person" or "the GTM lead" and those labels can get sticky.

What helps: Volunteer for cross-functional stretch projects. Use demos, reviews, and even LinkedIn posts to shape the narrative of your expanding range.

Conclusion: Your Career is a Landscape You Build

Finding your first zone of strength is how you start. Expanding across new regions is how you grow into your fullest potential. You don't have to get it perfect. You just have to keep moving forward intentionally choosing roles, teams, and challenges that open new doors rather than close them.

Remember: Your first step sets the tone; Your next steps shape the journey.

And every new region you master—every new layer you add—makes you not just a better GenAI product manager, but a builder of the AI future itself.

Now that you know how to choose your first region wisely, let's talk about how to actually *land* that first GenAI PM role.

Because knowing where you want to go is powerful. But learning how to open the right doors? That's the next part of the adventure.

Chapter 11: How to Secure an Interview for Gen AI PM Role

Let's start with a truth most people miss: You're probably closer than you think.

Not because you've mastered every GenAI concept. But because you're asking the right questions:

- *"Where do I fit in this GenAI world?"*
- *"What kind of role actually suits my strengths?"*
- *"What do I need to show to get the first conversation?"*

You're not just scrolling job boards. You're being intentional. You're mapping your strengths. You're reading this book. That already puts you ahead of most applicants.

Yes, the GenAI PM market is competitive—roles can attract hundreds of applicants per opening, especially in high-growth markets. But here's the twist: most of those applicants are guessing. They apply broadly with generic resumes, chasing credentials instead of clarity. The ones who stand out? They show signal—the ability to think like a GenAI PM, even before holding the title.

Picture this:

Sarah, a traditional product manager with a knack for stakeholder alignment, stumbles across a GenAI PM posting. She's intrigued—building AI-powered tools sounds like the kind of future she wants to help shape. But then she reads the requirements: "Model evaluation experience," "prompt engineering fluency," "track record of launching GenAI features." Her heart sinks. "I don't have that," she thinks. "I'm not the perfect candidate." So she closes the tab.

But here's the truth: Sarah was closer than she thought. That same role was filled by someone who didn't check every box either—but someone who told a clear story, showed curiosity, and proved they could learn.

The myth of the perfect candidate stops a lot of talented people before they even try. This chapter is here to help you break through that myth—not by pretending the road is easy, but by showing you how to walk it strategically.

And it's a path worth taking. As we noted in Chapter 2, global spending on AI is projected to hit $749 billion by 2028, with Generative AI alone accounting for over $200 billion, growing at a staggering 59% CAGR. Meanwhile, the 2025 LinkedIn Work Trend Index shows that AI skills are now the most in-demand capabilities across product, design, marketing, and business roles. Four in five global leaders say they're rethinking how their organizations operate—and they need talent that can help lead the shift.

So here's your real goal:

You're not aiming to be perfect. You're aiming to signal readiness—a clear, compelling story that shows you can think like a GenAI PM, even if you haven't held the title yet.

Maybe you've led complex launches as a core PM. Designed intuitive UX flows that built trust. Shipped APIs that required serious orchestration across systems. Or maybe you've driven innovation from outside tech—like redesigning a patient intake process, improving customer retention through better onboarding, or leading operational change in a regulated environment. These aren't side notes—they're your foundation.

In Chapter 10, we mapped your possible entry points into GenAI PM land—and explored how to find your starting zone. Now it's time to take the next step: opening the door to your first GenAI PM interview.

This chapter will walk you through that process—practically, honestly, and with momentum in mind.

Why GenAI PM Hiring Is Different

Now that you know the opportunity is real, let's talk about the process — because landing a GenAI PM interview isn't just about doing more of what worked in traditional PM job hunts. It's a fundamentally different game.

If you approach it like any other product role — polishing your resume, applying to job boards, waiting for a recruiter to call — you'll likely get stuck. Not because you're unqualified, but because the rules are still being written.

Chapter 10 was about choosing the *right* path. This chapter is about walking it— how to build proof, shape your narrative, and create momentum that turns intention into opportunity.

Here's how the GenAI hiring landscape actually works — and what it means for your strategy.

The Roles Are Still Evolving

Most GenAI PM roles don't come with clear roadmaps—not in the work, and not in the job description either.

You'll find listings that say "3+ years of experience with agentic orchestration" (a field that barely existed two years ago) or ask for both deep ML knowledge and strong design instincts. That's not a coherent brief—it's a wish list.

What does this mean for you:

You can't take job descriptions too literally. You need to read between the lines. What challenge is the team likely facing? Where are they confused? Can you position yourself as someone who brings clarity—even if you don't meet every bullet point?

The Work Is Ambiguous by Design

Unlike many traditional PM roles that emphasize delivery and scale, GenAI PMs are often hired to explore the unknown—testing early capabilities, prototyping fast, and adapting to rapidly shifting user behavior.

You're not just writing specs. You're shaping product behavior in systems that learn, iterate, and sometimes hallucinate.

"The best way to stand out in GenAI isn't to wait for clarity. It's to help *create* it."

What does this mean for you:

What hiring managers want isn't polish. It's adaptability. They'll be looking for signs that you're comfortable with messiness—that you've built v1s, worked with R&D, or made smart product calls even with incomplete data.

Signal > Resume

With so much noise in the market, hiring managers is increasingly filtering for signal—and they don't always find it in your job title or education. You don't need a GenAI title to prove you can think like a GenAI PM. What hiring managers

261

notice isn't credentials—it's signal. Tangible proof that you've already started developing product instincts for this space.

Here are three high-leverage ways to do that:

A smart teardown of a GenAI product on Medium.

A thoughtful series of GenAI product teardowns—each showing how a real feature works, where it falls short, and how you'd improve it from a PM lens.

You don't need to go viral or publish every week—but a couple of well-structured LinkedIn posts or Medium articles can show you're thinking rigorously about GenAI products in the wild.

Pick tools you've used (or studied), like Notion AI, GitHub Copilot, Miro's AI mapping, or Duolingo. Dissect the user flow, identify pain points or trust issues, and propose grounded improvements—especially in prompt strategies, UX scaffolding, or behavior tuning.

Example: "Why Notion AI sometimes overwhelms users—and how a simpler prompt layer could help." You walk through the current experience, identify where user expectations diverge from outputs, and propose a redesign using role-specific prompt modes.

Why it works: It's public, product-focused, and rooted in user empathy—all key GenAI PM instincts.

Note: Couple of teardowns like this—shared in public—helps position you as someone who thinks deeply about GenAI products. They don't just build your credibility online—they also give you strong material to reference in your resume or bring up in interviews as proof that you're a self-starter with real product instincts.

A crisp outreach message that reflects research

A short, personalized LinkedIn message to a hiring manager that names a real challenge their team is likely facing—and offers a thoughtful nudge or insight.

The key isn't just to pitch yourself. It's to reflect that you understand the team's product space—and are already thinking like someone who belongs there.

Example: "Hi Priya, I saw you're hiring for a PM on the [Healthcare Copilot] team. I've been exploring prompt testing frameworks to improve factual reliability in medical GenAI tools. Based on the product's current behavior, I imagine hallucination boundaries across patient-facing flows are a key challenge. I'd love to learn more or share a few ideas I've been prototyping."

Why it works: It's short. It's specific. It doesn't brag. And it offers value, rather than asking for favors.

A concrete resume bullet that shows product thinking

A resume bullet that shows how you shaped model behavior—through prompt design, user testing, or evaluation—even in a self-driven project. You don't need to have shipped this at work. A hackathon, side project, or prototype is enough—if framed the right way.

Example resume line: "Designed and tested prompt strategies for a GPT-powered learning assistant prototype—improved answer accuracy by 28% through iterative prompt tuning and user testing."

Why it works: It highlights experimentation, product focus, and real metrics—all in one line.

What does this mean for you: Focus on showing how you think, not just what you've done. Product instincts, cross-functional savvy, and AI curiosity speak louder than certificates.

No One's Waiting for You to Be 'Ready'

The tech is evolving too fast for anyone to be fully prepared. The PMs who break in? They don't wait. They prototype, contribute to early-stage initiatives, experiment in public, or get involved in GenAI-adjacent work inside their current org.

What does this mean for you:

Don't let lack of formal experience stop you. If you've built a working demo, led an AI-lite feature, or can articulate product trade-offs in this space—you're already more ready than most.

You're not applying to fill a box. You're applying to help shape what the box becomes. Hiring managers aren't looking for experts. They're looking for momentum. And momentum starts with how you position yourself—clearly, honestly, and with intent.

Position Yourself with Proof

Before you start applying, pause and take stock—not of what you lack, but what you bring.

You don't need to have shipped an agent or fine-tuned a model to land a GenAI PM interview. What matters is how you position what you have done—and how clearly you connect it to the kind of thinking GenAI product work demands.

Start with this question: *"What do I already do well—that maps to the needs of the specific GenAI PM roles I'm targeting?"*

If you've worked in tech, maybe you've launched complex products, scaled infrastructure, or built trust through design. But even if you come from healthcare, education, retail, or consulting—you've likely navigated ambiguity, managed risk, earned user trust, or driven innovation under constraint. These aren't detours. They're raw materials.

Your job is to connect the dots between that material and the qualities GenAI teams actually value:

- Comfort with ambiguity
- Systems thinking
- Fast learning loops
- Human-centered thinking
- Experimentation under uncertainty
- And above all, a *full-stack builder's mindset*

A 2024 LinkedIn survey found that 76% of product hiring managers prioritize transferable skills and curiosity over direct experience. That means your background can become a signal—*if* you help people see how it translates.

Here's what that can sound like:

- A developer might say: *"I'm used to debugging complex interactions between APIs and service—the same kind of pattern-mapping GenAI features require when connecting prompts, models, and tools."*
- A UX designer might say: *"I've spent years helping users trust opaque systems. Now I want to apply those skills to AI interfaces—where clarity and control are everything."*

- A business strategist might say: *"I've led pricing and positioning for emerging tech. GenAI is another frontier where understanding value, risk, and messaging deeply matters."*

These aren't pivots. They're logical extensions of what you already do well.

Here's the other key: don't pretend to know more than you do. You don't need to fake technical depth—but you *do* need to show that you're engaged and learning with purpose. That could look like:

- Building a small side project that uses GenAI APIs.
- Writing a teardown of a GenAI product you admire.
- Taking a course like *"AI for Product Managers"* or *"Generative AI for Everyone"*—and reflecting on what it taught you.
- Participated in an AI hackathon or local meetup.

These don't need to be big, flashy moves. What matters is that they show movement — proof that you're already investing in the space, not waiting on the sidelines. That alone sets you apart from most applicants.

So don't focus on what's missing from your resume. Shine a light on how you're building forward—and why it matters.

When you position yourself clearly, confidently, and truthfully, every application becomes more than a formality. It becomes a signal. Now let's translate that signal into action—starting with how to find the right opportunities and tailor your approach.

The ACTIONS Framework: From Intent to Interview

Positioning yourself with clarity is the first step. But how do you move from that strong narrative to actually securing conversations? That's where the ACTIONS framework (see Figure 11.1) comes in—a practical, momentum-building approach to getting your first GenAI PM interview.

Letter	What It Stands For	Key Question It Answers	Real-World Step	Outcome by End of This Step
A	Assess Opportunities Smartly	Where should I play?	Identify a focused shortlist of GenAI PM roles (internal or external) that align with your strengths, learning goals, and long-term momentum	A list of high-signal roles you're intentionally targeting
C	Consider Skills and Gaps	Am I ready for this role?	Compare your strengths and experience against your target roles — identify critical knowledge, exposure, or credibility gaps	A clear map of gaps to close and strengths to highlight
T	Talk to Mentors	What am I missing — or not seeing?	Share your target roles and gap map with trusted PMs, engineers, or mentors to validate your thinking or course-correct early	Informed input that builds confidence — or helps you pivot
I	Improve Qualifications	How do I close the gap?	Fill high-leverage gaps by joining AI pilots, taking a course, writing a teardown, or launching a side project	Fresh, credible proof points that elevate your profile
O	Optimize Resume and Signal	How do I present myself credibly?	Tailor your resume, update your LinkedIn, write or reshare public proof (e.g., GenAI teardown, small prototype, article)	Polished application assets that reflect your unique story and readiness
N	Network and Apply Intentionally	How do I get noticed and referred?	Apply with purpose — reach out to PMs, hiring managers, or connectors with context and curiosity; submit sharp, role-specific applications	One or more applications in active review; warm leads and ongoing conversations
S	Secure the Interview	How do I turn interest into interviews?	Convert relationships and applications into actual interviews — by staying early, relevant, and remembered	A confirmed interview (or screen) with one or more GenAI PM roles

Figure 11.1 The ACTIONS framework for landing a GenAI PM Interview.
© 2025 by the author of *Zero to GenAI Product Leader*.

Each letter represents a step you can take—strategically, not blindly—to go from *curious* to *considered*. Figure 11.1 provides a detailed breakdown of each step, including the key question it answers, the real-world action to take, and the outcome to expect. Let's dive into these steps to turn your intent into action.

A—Assess Opportunities Smartly

Ask: Where should I play?

Before you hit "apply," pause.

Not all GenAI PM roles are created equal. Some will stretch you in the right ways. Others will burn time without building traction. And in a space where even the job descriptions are still evolving, the biggest risk isn't being underqualified—it's being undirected.

This step is about finding direction. Your goal here isn't to apply. It's to map the landscape—and identify a shortlist of high-leverage roles worth pursuing. These roles won't just check a box on your résumé. They'll build your instincts, increase your exposure to real GenAI systems, and unlock momentum for what comes next.

You'll revisit and refine your role list later in Letter O. But right now, your job is to zoom out and ask:

"Where can I grow fast, contribute credibly, and learn on the front lines of GenAI?"

Start where you already have traction

If you're already working inside a company that's investing in AI, look inward before you look outward. Internal transitions are often the lowest-friction, highest-leverage way to step into GenAI work.

You have trust. You have context. You know the stack, the stakeholders, and the politics. That gives you an edge no resume can replicate.

Sometimes, the most strategic move isn't jumping ship—it's raising your hand early. Join an internal pilot. Volunteer for a GenAI task force. Offer to partner on evaluation testing or prompt design. Even informal exposure builds credibility, fast.

A PM I recently coached moved into an LLM evaluation role simply by attending an internal hackathon—and following up with the right engineer afterward. No job board. No interview loop. Just momentum.

And if you look outward—Do it with precision

If your current company isn't serious about GenAI—or you're ready for a new challenge—it's fair to look outside. But don't cast a wide net. Every external application costs time, attention, and emotional energy. Tailored resumes. Company research. Thoughtful outreach. It all adds up.

So instead of asking, "Where are the jobs?" ask:

"Where can I credibly contribute—and learn at speed?"

Maybe that's a mid-stage SaaS company embedding AI into its core workflows. Maybe it's a platform team launching orchestration layers across models—like what you'll see in Azure AI Foundry. Or maybe it's a startup building out agents, where product thinking needs to evolve alongside the architecture.

What matters most isn't the label—it's the learning surface.

How to know a role is worth pursuing

There's no perfect checklist. But here's what you're really scanning for—beneath the titles and buzzwords:

- Will I be working on a core GenAI challenge—shaping behavior, enabling usage, or driving adoption at scale?
- Are PMs empowered to make product decisions—or mostly executing someone else's vision?
- Will this role help me build meaningful experience—or just keep me in motion?

Look for roles that give you front-row access to how GenAI systems behave, adapt, and drive outcomes—whether that's through user interaction, infrastructure orchestration, ecosystem distribution, or agentic workflows that chain tools and reason across steps. If the role plugs you into a real feedback loop—where your work shapes model behavior, user trust, or system performance—it's worth leaning in, even if the title isn't flashy.

If you've shipped 0 to 1 product, navigated ambiguity, or led sensitive user-facing experiences—you're already working with the kind of instincts GenAI teams need. The trick is framing that clearly. These experiences are often more transferable than they look on paper.

If you're early in your career—or pivoting in from UX, operations, or data—look for roles that place you near the learning loop, even if they don't come with a bold PM title. Associate PMs, AI product analysts, or evaluation-focused positions often pair you with senior PMs, giving you a structured path to contribute while building fluency. The best early roles aren't always loud—they're the ones that let you learn in public, with a net.

And if you're a domain expert—in healthcare, legal, finance, or education—that's not a limitation. It's leverage. As more companies build vertical-specific copilots and agents, they're looking for PMs who understand those workflows deeply. Even if your AI depth is still growing, your domain fluency might be exactly what makes you stand out.

Shifts toward multi-model strategies and agentic systems—which we've discussed throughout this book—are already changing what GenAI PMs are expected to handle. So as you scan new roles, look for environments that stretch your systems thinking, not just your resume.

Energy matters more than ever

The GenAI market moves fast, but interviews don't. Application loops can stretch across weeks. Assessments vary wildly. Some teams expect a product teardown. Others want a prompt strategy doc. Some don't know what they want—yet.

That's why your time and energy are your scarcest resources. Protect them. Don't apply reactively. Apply surgically. A short list of 5 roles where you have a story—and a plan—will beat a scattershot of 25 generic submissions every time.

Where to start looking—and how to look smartly

If you're searching beyond your current company, the problem isn't scarcity—it's noise. GenAI Product Manager roles are popping up everywhere. But only a small fraction will truly sharpen your instincts or give you meaningful exposure to how GenAI is built.

So don't just ask *"Where are the jobs?"* Ask: *"Where can I grow the fastest—and actually contribute?"*

The best roles often won't say "GenAI PM" in the title. Some are posted quietly as "Product Manager, Platform" or "LLM Tooling PM" on a model provider's careers page. Others surface in a founder's LinkedIn post, or a tweet from an engineer at LangChain saying "we're hiring."

That's because in GenAI, hiring often starts with builders—not recruiters. So your job isn't to scroll faster. It's to read smarter. Pay attention to patterns:

Are they building with open-weight models like Llama, Mistral, or Command R? Are they working on orchestration layers, developer tools, or multi-agent systems—not just shiny UX? Are they shipping fast, sharing trade-offs, and talking openly about model behavior?

If the hiring manager is active in the community—posting demos, discussing roadmap challenges, or contributing to open source—you've found a lead worth following. Don't cold-pitch. Show up informed. Ask a thoughtful question. Join the conversation before you need anything.

Some of the strongest roles never hit job boards. They begin with a teardown you posted, a prototype you built, or a smart comment in a Slack thread. That's why finding the right role isn't just about looking—it's about being visible *where the work is happening.*

Where the smart search happens
Start with LinkedIn—but use it intentionally.

LinkedIn's AI-powered job search lets you describe the kind of role you want in natural language. Try: "Product Manager building GenAI APIs for developers" or "PM for multi-model orchestration systems."

Filter by *Remote*, *Past Week*, or *Contract* based on your flexibility. Follow PMs, engineers, and researchers from model providers or infra teams—they often post jobs before recruiters even get involved.

Next: explore startup hubs.

Wellfound (formerly AngelList) remains a goldmine for early-stage teams. Use filters like *AI/ML*, *Remote*, and *Product*. Look for startups hiring their first or second PM—they offer steep learning curves and a front-row seat to shipping fast. But remember these teams won't slow down to teach you GenAI. They'll expect you to contribute fast, learn on the fly, and shape product decisions from day one.

Here's a pro move: track funding rounds. Most startups start hiring 2–6 weeks after raising funds in the next round. Use Crunchbase, StrictlyVC, or newsletters like *Not Boring* (by Packy McCormick) to stay ahead of the next wave. These sources often spotlight startups right after they raise—which is when hiring quietly begins.

Follow builders, not just brands.

Open-source contributors, product leads, and engineers at companies like Hugging Face, Anthropic, Cohere, AWS Bedrock, Google Vertex AI or Microsoft's Azure AI often announce roles first on X (Twitter), LinkedIn, or in Slack communities.

Join niche spaces like *AI Product Hive*, *Women in AI Product*, or *Latinas in AI*—where hiring managers casually share leads, long before the JD is finalized.

What if you don't find "The One" right away?

Let's say none of the titles feel quite right. Maybe you're not seeing a perfect GenAI PM fit—or the roles you want feel just out of reach.

That's not a dead end. It's a signal to zoom out. Because the smartest move isn't always finding your dream role on day one. It's the one that puts you closer to the action—where you can learn, contribute, and build momentum. Let's talk about that next.

Play the long game

Let's be real—even with smart strategy and clear signal, you might not land your ideal GenAI PM role on the first try.

That doesn't mean you're off-track. Some of the most effective PMs in GenAI didn't start with a perfect title. They found a way to get close to the action—by stepping into roles that built proximity, relationships, and trust. Then they earned their way forward.

If you're not landing PM interviews yet, don't panic. Step one is getting in the room. That could mean taking on an adjacent role that gives you access to real GenAI behavior—like prompt scaffolding, tool orchestration, or evaluation work—even if the job doesn't say "product" in bold letters.

These roles might include:

- Technical Program Manager on an LLM infrastructure team
- AI Solutions Engineer working directly with customers and GenAI APIs
- Research Ops or Evaluation Lead for model teams
- Strategy or BizOps roles focused on AI monetization or go-to-market

No, they won't build your PM portfolio overnight. But they'll teach you how these systems behave — and put you in the circles where product decisions are actually being made. When the next PM role opens up? You'll already be there. Already contributing. Already someone they trust to step up.

I've seen this up close. My own transition into product didn't start with a formal application. I was in marketing, collaborating with a PM on a high-stakes launch. When he moved on, he suggested I replace him. Not because I had the résumé— but because I'd built trust, asked good questions, and shown I could do the work.

The lesson? Don't fixate on the title. If a real opportunity presents itself—even if it's not perfect or exactly what you hoped for—consider taking it. Don't wait for the dream door. Step through the real one.

Why this isn't the moment to apply yet

You'll apply later—after you've refined your story, built signal, and tailored your résumé. That's what Letters C through O are for.

But by the end of this step, you should already have: A list of 5–10 promising roles or teams, a sense of where your experience could map to their goals and a few people in mind who might help you test your thinking

This isn't about rushing into action. It's about choosing where to act—and where to invest your time next.

Next: From direction to readiness

You've mapped where you want to play—the roles that stretch your instincts, not just your résumé. But clarity alone doesn't secure interviews. Now it's time to get honest about what it'll take to be seen as a credible candidate.

What's missing—in knowledge, exposure, or proof? Where are you already strong—and where do you need to grow fast? That's what we'll unpack next.

C—Consider Skills and Gaps

Ask: Am I ready for this role?

Now that you've got a shortlist of high-signal roles, it's time to get real: What will it take to be competitive for these roles—not someday, but now? This step is about clarity, not self-doubt.

You don't need to be perfect. But you do need to be prepared. The goal here isn't to check every box—it's to know which ones you already cover, which ones you can learn fast, and which gaps might trip you up if left unaddressed.

That's how you build a plan—not from fear, but from focus. This is where clarity turns into traction.

Look at your shortlist like a hiring manager

Pick 2–3 roles from your list. Read them slowly. Don't get caught up in jargon. Instead, ask:

- What are the top 2–3 capabilities they're really asking for?
- Look beyond the title—and into the work. What kinds of product decisions might this team be making? What problems are they likely trying to solve?
- What would I want someone in this role to already know—or be eager to learn?

Then do a soft audit of your own background. You're not grading yourself—you're mapping terrain.

Maybe you've collaborated closely with engineers or analysts—even if not in a formal product role. Maybe you've scoped workflows, coordinated launches, or shipped internal tools in ambiguous environments. Maybe you've built credibility in a domain like healthcare, legal, education, or finance—where GenAI adoption is accelerating.

That's all signal. And the places where you feel unsure? That's not a red flag. It's a guidepost for what to focus on next. The trick isn't to over-index on buzzwords but to understand the shape of the job—and whether you can grow into it quickly.

Gaps come in three types—know which one you're dealing with

Knowledge gaps: You don't fully understand what's being discussed—e.g., prompt engineering, model routing, vector databases. That's okay. You can learn them fast. You can close the gap through short courses, tutorials, or even shadowing others if that's an option.

Exposure gaps: You haven't actually worked with these tools or concepts yet—but you get them conceptually. This is where side projects, teardown posts, or AI hackathons come in. A little exposure goes a long way.

Credibility gaps: You've done adjacent work—but it's not obvious to others how it translates. These require storytelling. You can bridge these gaps through framing:

reframing your past experience to show how it maps to GenAI work. The idea is to make it easy for the hiring team to understand how your experience ties to the role.

Most people mix these up. They take a credibility gap (e.g., "I've never worked with GenAI tools in a PM role") and mistake it for a knowledge or capability gap ("I'm not qualified"). That's how people talk themselves out of trying. Don't do that. Your job isn't to fix every gap—it's to know which ones matter, and how you'll address them.

Quick tip: Use a job description as a diagnostic

Pick one of your targets GenAI PM roles. Paste it into a doc. Then go line by line and write: I've done this or I've done something similar or I haven't done this yet.

Now look at "I haven't done this yet" list and ask: Are these coachable in 30–60 days? Can I demonstrate learning potential here, even without direct experience? If the answer is yes, that's not a blocker—that's an opportunity to show growth.

You're not starting from scratch. You're starting from strength. The question isn't *"Do I belong?"* It's *"What story do I need to tell — and what gap do I need to close — to get in the room?"* That's what we'll answer in the next steps. But for now, you're building a map.

A map of where you stand. A map of what matters next. A map that will guide how you talk to others, how you learn, and how you show up.

Note: You can even use GenAI tools like ChatGPT, Grok, or Gemini to assess your readiness. Paste in your résumé and a job description, then ask the model to compare the two:

- *What looks like a strong match?*
- *What feels like a gap — and is it a must-have or nice-to-have?*
- *How might you reframe your experience to better align?*

These tools can also help you close knowledge gaps fast — or even help you frame your credibility gaps in a way that resonates with the role. Use them as a thinking partner — not just for answers, but for clarity.

Let's move to T—Talk to Mentors—because sometimes what you need most isn't another checklist. It's a fresh set of eyes.

T—Test Your Thinking

Ask: What am I missing—or not seeing?

You've mapped the terrain. You've scanned the job descriptions. You've even taken a first pass at your own strengths and gaps. But even the sharpest self-audit has blind spots. That's where this step comes in.

Before you apply—pause. And test your thinking out loud with someone you know. This isn't about validation. It's about calibration. You're not looking for someone to say "Yes, go for it." You're looking for someone who'll say, "Here's how you could go at it better." Don't ask, "Can I do this?" Instead ask, "Where would you place me—and why?" That single shift turns a dead-end question into an open door—for insight, for feedback, for realignment.

The best sounding boards—whether inside or outside your field—won't just say yes or no. They'll help you:

- Spot skills you're underselling
- Reframe your experience in more relevant language
- Flag roles that sound promising but may burn your energy
- Suggest shortcuts you hadn't considered—like internal projects, referrals, or shadowing opportunities

This isn't about validation. It's about calibration. You're testing your thinking *before* you hit "apply."

Who should you talk to?

You don't need a big-name mentor. You need someone who knows the game—or knows your game. That might be:

- A PM or engineer already close to GenAI work, who can help you spot where your current skills overlap—and where they don't.
- A former peer or college alum who's already made the leap, and remembers what it's like to be standing at the edge.
- A hiring manager who's seen hundreds of résumés and knows what truly stands out—and what doesn't.
- A senior stakeholder in your org who knows your track record and can point you toward internal opportunities—or sponsor your transition.

- A second-degree LinkedIn connection, someone you can reach via a warm intro or shared context—like a Slack group, meetup, or AMA.
- A speaker you admired at a panel, podcast, or event. Reference their work. Ask a sharp follow-up. Let curiosity lead.

And don't forget the most underrated source of clarity:

A friend, partner, or former manager who knows how you think—and isn't afraid to ask the questions no one else will. Some of the most clarifying feedback doesn't come from someone in tech—it comes from someone who knows you.

My partner's a school teacher. When I explained a GenAI role I was eyeing, she asked, *"Okay, but what would you actually do in that role?"* That one question reshaped how I thought about the fit.

You're not seeking answers. You're seeking sharper questions. That's where clarity begins.

What should you share?

Come prepared. You'll get better advice if you give better inputs. Basically, make it easy for someone to help you. Share a tight summary of where you're headed—and ask them to pressure-test it. One script that works well:

"I've been exploring GenAI PM roles and have narrowed it down to three paths that feel aligned with my background and learning goals. Based on your experience, I'd love your take on a few things:

- *Where do you think I'm best positioned to contribute today?*
- *Are there any areas I might be underestimating — or overreaching?*
- *And what would you suggest I focus on in the next 30–60 days to sharpen my odds?"*

You can also share a rough résumé, a skills map, or even your draft 2-minute story for "Why GenAI PM, why this role, why now?" Let them challenge the framing—and help you tighten the story. Clarity often sharpens when it's said out loud.

Don't force the next step

You're not asking for a referral. You're building insight. If the conversation goes well and the fit is real, people will often volunteer support—whether that's a second conversation, an introduction, or a heads-up about a role.

If you want to gently leave the door open, try:

"Thanks again for the feedback—it's incredibly helpful. If something opens up that feels like a good match, I'd love to stay in touch."

No pressure. Just presence. You're not looking for approval. You're seeking a signal. The best mentors won't say, "You're ready." They'll say: *"Here's what I'd do next if I were you."* That's all you need to keep moving.

Let's head to I—Improve Qualifications because insight without action won't close the gap. Let's talk about how to build just enough proof to make the shortlist.

I—Improve Qualifications

Ask: How do I close the gap?

At this point, you've done what most jobseekers skip. You've narrowed your search. You've assessed what's missing. You've even pressure-tested your plan with who knows the game—or knows your game.

Now it's time to do the work. But not all effort is equal—and not every gap needs a bootcamp. This step is about targeted moves that build *just-in-time fluency*—enough to speak the language, show initiative, and demonstrate you're serious.

The good news? In GenAI, signal compounds fast.

A single, well-crafted effort—like a sharp teardown, a lightweight prototype, or a thoughtful side project—can shift how others see you: from curious onlooker to emerging contributor. You don't need fancy tools or months of prep. Just clarity and execution. Maybe you spend a weekend analyzing a product like Claude, draft three specific suggestions in a five-slide deck and post it on LinkedIn with a smart caption. That kind of work shows you can think like a PM—spotting friction, proposing tradeoffs, and packaging ideas clearly.

Done well, it won't just showcase interest—it'll spark feedback, conversation, and sometimes, a DM from someone hiring. In startups, it might lead to an exploratory chat or a warm intro. In bigger companies, it sharpens your narrative and gives you tangible proof to reference during interviews.

This is the builder's mindset in action—not just learning passively, but shipping something small, smart, and visible.

No, it's not a golden ticket. A sloppy post can backfire. But a tight, targeted proof point—paired with a credible application—can tip the balance. Especially in a space where initiative often speaks louder than pedigree.

Focus on the smallest move that builds real proof

You don't need to spin up a full-stack app or enroll in a 10-week course. You need to fill the right gap, the fastest way.

Let's say you're missing hands-on exposure to model evaluation. Could you shadow an internal prompt audit? Could you run a mini experiment using OpenAI's playground or Azure AI Foundry Playground or Hugging Face datasets—and document what you learned?

Or maybe you're strong on the product side, but light on AI concepts or coding. Start small. Try a teardown of a GenAI product you admire—like Microsoft Copilot or Claude—and break down what works, what's confusing, and what you'd improve. A five-slide deck shared on LinkedIn or X can show sharp product thinking in a space that moves fast. You can also mockup a user flow for a retrieval-augmented feature using Figma or Notion—especially if you want to demonstrate UX fluency around GenAI concepts without needing to code.

If you're technically curious, tools like Glide, Streamlit, or ChatGPT's custom GPTs let you simulate workflows or build lightweight AI demos without deep engineering knowledge. These are realistic weekend projects—often doable in under 10 hours—that let you explore user needs, prompt design, and AI interaction loops.

When you're ready to go deeper, frameworks like LangChain or Semantic Kernel can help you explore orchestration and tool use—but only if you're comfortable with code or have someone to collaborate with, like a coder friend you can shadow.

The point isn't to build something perfect. The point is to build momentum—and a story you can point to when someone says, "Tell me what you've done in GenAI so far." A polished, bite-sized project shows you're not just interested—you're already thinking like a GenAI PM.

Pick your fastest visible win

Different gaps call for different moves. Here are a few high-impact ways people build credibility fast:

- Write a teardown. Analyze a GenAI feature you admire. What trade-offs did they make? What would you improve? Post it on LinkedIn or Medium.
- Join an AI hackathon. Sites like Devpost or HackerEarth host them regularly. Even just participating gives you exposure, teamwork stories, and a potential repo.
- Contribute to a GenAI side project. Can you help a friend, nonprofit, or startup prototype something using AI APIs? You don't need to lead it. Just contribute with intent.
- Enroll in a focused AI course. Not everything needs a certificate—but 10–15 hours on Coursera, DeepLearning.ai, or Product Academy can fill key knowledge gaps.
- Document your learnings. Don't just watch a video. Reflect on it. Write your take. This builds recall *and* public proof.

Note: Pair *learning* with *signal*. Every time you learn something, ask: *"Can I turn this into a tangible proof point?"* A post. A prototype. A GitHub readme. That's how you get noticed — even while you're still learning.

You don't need to be ahead of the curve—just moving

Hiring managers don't expect you to know everything. But they absolutely notice when someone's closing the gap and taking initiative without waiting for permission.

One product marketer I worked with had zero formal AI experience—but spent a few weekends experimenting with OpenAI APIs, wrote a LinkedIn post about building a small customer-support demo, and shared her learnings with two engineers she'd met at a local product meetup. That side project and follow-through opened up a warm intro to a hiring PM—and led to her first GenAI interview.

It wasn't because she was "qualified." It was because she was already building.

Keep it scrappy—but intentional. You don't need more projects. You need one or two moves that send a clear message:

"I know where I want to go—and I'm already in motion."

Next up: You've built the proof. Now it's time to package it. Let's move to O—Optimize Resume and Signal—where we turn all of this momentum into assets that get noticed.

O—Optimize Resume and Signal

Ask: How do I present myself credibly?

You've clarified your story. You've identified and closed key gaps. You've even had some early conversations to pressure-test your thinking.

Now it's time to make that progress visible. This step is about two things:

- Refining your target role list based on the latest openings
- Sharpening your résumé and public proof so your story travels ahead of you

Before you apply, update your role list. Go back to the roles you shortlisted in Step A. Some might be closed or already filled. Others may no longer feel like a strong fit—or new ones may have popped up that are even better.

Do a fresh sweep of: LinkedIn, Company career pages, AI-specific job boards like ai-jobs.net or Wellfound

And this time, don't just look at the job title. Reread the description with fresh eyes—now that you've refined your story and added signal. You might spot stronger fits than you did earlier.

You're not starting from scratch. You're updating your map.

Step 1: Tailor your résumé to the role—not just your past

Most résumés read like a personal archive. Yours should read like a product pitch: "This is the person you're hiring for this role."

Start with one of your target roles open in front of you. Then ask:

- What's the through-line between what they're hiring for — and what I've done?

- What phrases from the JD can I naturally echo in my bullet points?
- Where can I quantify outcomes—or clarify scope—to make my impact sharper?

If you've shipped internal tools, worked on search, chat, or automation—make those visible. If you've built products under uncertainty, collaborated with ML teams, or navigated data tradeoffs—make that obvious. These are all skills GenAI PMs need.

And don't over-explain GenAI basics. Assume the reader understands the space — but doesn't know *you* yet.

Note: Want to go deeper on how to tailor your resume—line by line—using AI tools like ChatGPT? Check the bonus section at the end of this chapter: How to Tailor Your Resume for GenAI PM Roles.

It includes formatting best practices, ATS tips, and example prompts to use AI tools for gap analysis, bullet point polishing, and more.

Step 2: Align your LinkedIn—and let it work for you

Your LinkedIn isn't just a résumé. It's your narrative anchor.

Make sure:

- Your headline reflects what you're moving toward (e.g., "Product Manager | Building AI tools for developers" vs "PM at [Company]")
- Your summary tells your GenAI story: where you're strong, what you've built or explored, what you're looking for next
- Your experience section mirrors the sharpest proof points from your résumé — with keywords that will pass AI recruiter filters (e.g., "LLM," "MLOps," "prompt tuning," "agent orchestration")

Also: follow companies and PMs in your target space. Engage occasionally. Many hiring leads check your activity—and a visible presence helps them remember you.

Step 3: Make your proof easy to find—and easy to trust

By now, you've probably built or written something that shows your learning in motion. That's your proof. Now it's time to make sure it actually reaches the people who matter.

That doesn't mean shouting. It means shaping:

- Update your résumé to reflect what you built—even if it was self-initiated. Use metrics, clear descriptions, and outcomes where possible.
- Link to your teardown, prototype, or write up at least in your *LinkedIn Featured* section.
- If it's a Notion page or GitHub repo, include a short framing paragraph— why you did this, what you learned, and what it shows about how you think.

Your goal isn't just to have proof. It's to make it legible and relevant to the person scanning your story for the first time. Because you've already done the work. Now it needs to work for you. One quick test, Look at your LinkedIn and résumé side by side. Then ask:

"Would someone believe I'm ready to interview for a GenAI PM role—without a backchannel conversation?"

If the answer is "not quite," you know where to focus next. You've done the work. Now make it legible. Because when someone scans your profile or your doc, they shouldn't have to guess:

- What have you done?
- What are you aiming for?
- Why are you fit?

You're not just being evaluated on your experience. You're being evaluated on your clarity. Let's move to N—Nudge the Right Conversations—because even the sharpest résumé won't move on its own. You need to get it in front of people who matter.

N—Nudge the Right Conversations

Ask: How do I get noticed and referred?

You've done the prep. You've shortlisted a handful of roles, built a credible narrative, created signal, and tailored your résumé. Now you're ready to apply—and you should. Great roles fill fast, and your résumé deserves to be in that first wave.

It's tempting to wait—to hold off until you've spoken to someone or secured a referral. But here's the truth: just applying may not land you the interview. Yet not applying could mean missing the window altogether. Think about it from the recruiter's perspective.

Most GenAI PM openings attract hundreds of applicants. And while a few are reviewed thoughtfully, many are triaged by filters or skimmed at speed. If you're not in that first batch—often the top 20 applicants—your odds of organic visibility drop fast. That's why early application matters. It earns you a spot in the pile—and earns you the right to follow up. But submitting isn't enough. Because the strongest candidates don't just apply and wait. They pair the application with a nudge—a direct touchpoint with someone close to the role. A hiring manager. A recruiter. A teammate. A connector.

This step is about sparking the right conversations—with clarity, context, and credibility. Not with cold blasts or "please refer me" asks. But by showing up like a PM would: informed, intentional, and bringing signal that makes people say, "I want this person in the loop."

And who you reach out to—and how—depends on your relationship, the role, and the moment. Let's walk through those scenarios next.

Let's say you've identified a GenAI PM role X at Company Y. Here's how to approach each path.

You're reaching out to the hiring manager

This is the highest signal route—and the most delicate. If they're actively hiring, they're likely overwhelmed. So, lead with relevance and respect.

Here's how:

"Hi [Name], I saw your team is hiring for [GenAI PM Role]. I've worked on [brief relevant experience], and recently [shared teardown / shipped AI feature / led relevant project]. I'm genuinely excited about the work your team is doing, especially [specific detail about the role or product].

I've already applied, but I'd love to hear your take — and would appreciate any additional context or next steps you'd recommend. Here's my résumé and a short overview of what I'd bring to the role. Thanks so much for your time."

Why it works: You're showing initiative, not entitlement. You're asking for insight, not a handout. You're proving you've done the work

You're reaching out to the recruiter for the role

Recruiters don't need a full recap of your résumé—they need a signal and fit. Keep it crisp, relevant, and easy to forward.

"Hi [Name], I applied for the [GenAI PM Role] on [Team] at Company Y and wanted to check if it's possible to connect for 30 mins? I've recently [done X] and [built Y] — both of which align closely with what the role outlines. Here's my résumé and a quick summary of what I bring. I'd be happy to provide anything else needed. Looking forward to hearing from you. Thank you for your time!"

Note: If the recruiter hasn't posted the job themselves, mention where you saw the role and who's on the team — it helps them triage more quickly.

You're reaching out to someone in your network who works at company

This is one of the most strategic plays — especially if they're not the hiring manager but know the team. Your ask here depends on how warm the relationship is.

If they know you reasonably well:

"Hi [Name], I saw Company Y is hiring for a GenAI PM on [Team]. I've been focused on [X] and recently [shared teardown / built project] that aligns with what the team is building. I'd love to get your perspective on the role, or how the team's thinking about this space—and if it seems like a fit, I'd be grateful if you'd consider referring me or pointing me toward the hiring manager. Attaching my résumé and a short summary of what I'd bring. Appreciate you either way!"

If the relationship is cooler (e.g., met at an event, college alum, 2nd-degree contact), go one step at a time:

"Hi [Name], I saw your team is hiring for [Role] at Company Y and wanted to reach out. I'm currently focused on breaking into GenAI PM roles and have been working on [X] that feels closely aligned. I'd love to get your take on the team or the role — and if you think it makes sense, would really appreciate being connected to the right person. Totally understand if that's not possible—even a bit of guidance would be super helpful."

Remember: A conversation is often more valuable than a direct referral. Ask for input first—let them opt into the referral if the fit feels right

You're reaching out to someone who's not at company Y—But is connected to the hiring manager or team

Your goal here is to get warm intro. The ask is bigger, so set the stage with a short call or message explaining why you're reaching out.

If you're close:

"Hi [Name], I noticed you're connected to [Hiring Manager] at Company Y—I'm applying for a GenAI PM role there and think it's a strong fit based on [X]. Would you be open to making an intro? I can share a short note if helpful—and totally understand if not possible."

If you're not sure:

"Hi [Name], I've been exploring GenAI PM roles and saw a strong fit at Company Y. I noticed you're connected to [Hiring Manager], and would love to share more context and get your take on whether an intro might make sense. Would you be open to a 10-minute chat? I'd really appreciate your perspective."

You're reaching out to someone at company Y—But you're not sure If they're on the hiring team

In this case, your primary goal is learning—and *maybe* getting steered toward the right person. Here's how:

"Hi [Name], I saw you're at Company Y—and noticed a GenAI PM opening on [Team]. I'm deeply interested in the space, and recently [built/shared X]. Would love to hear your perspective on the team or the role—and if there's someone you'd recommend connecting with, I'd be incredibly grateful."

Keep it focused, not assumptive. They're not obligated to respond—but if they do, you're no longer cold.

Make your nudges count

This isn't about flooding inboxes. It's about showing up—like a PM would—with clarity, curiosity, and something to offer. When done right, nudges aren't interruptions. They're the beginning of trust.

So don't wait. Apply first. GenAI PM roles move fast—and your application earns you the right to follow up. Once you've submitted, layer in a thoughtful nudge: a short, relevant message that includes your résumé, a quick proof point, and a reason you're reaching out. No fluff. No friction.

Then, treat your outreach like a product funnel. Track who you reached out to, what you shared or asked, whether they responded, and if the conversation led to a follow-up, an intro, or a referral. Over time, patterns will emerge—and so will insights into how to sharpen your approach.

At this stage, you're not just applying. You're building signal density. You're making it easy for the right people to say, "I want to learn more." You're not waiting to be discovered. You're being remembered—for showing up with intention.

Next: S—Secure the Interview. Once you've put yourself on the radar, it's time to turn interest into action—and get that first call. Let's talk about how to close the loop and land the interview.

S—Secure the Interview

Ask: How do I turn interest into an interview?

You've done the work. You clarified your story, built proof, and reached out with intent. And now—a hiring manager's replied. A recruiter's pinged you. A referral has landed. This is where passive interest becomes active opportunity. But interest isn't the same as an interview. It's a door cracking open—and what you do next determines whether it swings wide or quietly closes. Here's how to make the most of that moment:

Step 1: Respond with confidence and clarity

When someone expresses interest—whether it's a DM reply, a recruiter message, or a response to your post—don't sit on it. Reply within 24 hours if you can. Momentum matters. Keep it light but intentional. For example:

"Thanks so much for the note—really appreciate it. I've been following [Company or Product] and love what you're building, especially around [area you care about]. Would be great to learn more about where this role is focused and how I might contribute."

You're not just saying "thanks." You're reinforcing that this isn't random—you've been paying attention, and you're here with purpose.

286

Step 2: Make the next move easy

Assume the person on the other side is busy. Reduce friction. You don't need to re-pitch your background (you already did that in your outreach or résumé), but a quick recap keeps context fresh and moves the thread forward:

"Quick background: I'm a PM with experience shipping 0 to 1 features in data-heavy tools, and I've recently been focused on GenAI workflows like prompt orchestration and evaluation. I've attached my résumé here again in case helpful."

And then suggest a clear path forward:

"Happy to share more context or hop on a quick call if useful. Let me know what works best on your end."

Simple. Confident. Easy to say yes to.

Step 3: Follow up—lightly, but deliberately

If you haven't heard back in 5–7 days, it's okay to follow up—especially if the conversation had positive early signals. Keep it light and respectful:

"Just wanted to circle back in case this slipped through. Still very interested—and happy to move at your pace."

If the response is "we're not moving forward," thank them, ask if they'd be open to staying in touch, and move on. You're playing the long game.

Step 4: When multiple threads open—prioritize wisely

Sometimes, momentum comes all at once—two or three promising threads start moving in parallel. Don't feel obligated to chase every one. Return to your filters:

- Does this role stretch my product instincts in the right way?
- Is the team building something meaningful—or just reacting to buzz?
- Will I grow fast here—or spend time convincing others why AI matters?

You didn't spray and pray. You aimed with intent. Now trust that instinct.

You're at the edge of the breakthrough

You've narrowed your focus, built real signal, and created meaningful conversations. At this point, you've either landed your first GenAI PM interview—or you're just one thoughtful follow-up away. But before we shift into interview prep, there's one more lever to sharpen your edge: Your résumé. Not the generic one—the one tuned to this space, this moment, and this role.

In the bonus section that follows, we'll walk through how to tailor your résumé for GenAI PM roles—and how AI tools can help you highlight what really matters. Once that's dialed in, you'll be ready for what comes next: Not pitching potential—but proving it. Let's finish strong.

How to Tailor Your Résumé for GenAI PM Roles—Using AI Tools

In a competitive job market, especially for specialized roles like Generative AI Product Manager, your résumé is often the first impression you make. And increasingly, it's not just a recruiter reading it—it's an AI. That's why your résumé needs to do two jobs at once:

- Pass the filters of Applicant Tracking Systems (ATS)
- Resonate with the actual hiring manager

This section walks through:

- Essential structure and formatting tips
- A proven framework to write better bullet points
- How to use AI tools like ChatGPT and Jobscan to sharpen and personalize your résumé

Start With the Right Structure

Recruiters spend an average of 30–60 seconds on a first scan. Your layout needs to make your story obvious—fast.

Sections that matter:

1. Contact Info: Clear name, email, LinkedIn link at the top. Keep it clean and professional.
2. Summary (Optional): If you're mid-career or pivoting into GenAI, a 2–3 sentence summary can help bridge your story. Focus on credibility + curiosity. If you're early in your career, skip it.

3. Education: List degrees, certifications, or bootcamps relevant to AI or product. Include institution, degree, and graduation year (optional if you're experienced).
4. Experience (Reverse Chronological):
   o Focus on impact, not just responsibilities.
   o Use bullet points (see formula below).
      Tailor each entry to reflect GenAI-relevant skills (e.g., experimentation, data work, trust-sensitive design).
   o Quantify wherever possible.
5. Skills (Optional): List hard skills (e.g., Prompt Engineering, LangChain, Python, Hugging Face) + soft skills (e.g., ambiguity, cross-functional leadership). Tailor these based on the role you're applying for.

Formatting Tips:

   o One page unless you have over 10 years of experience
   o Consistent fonts, alignment, spacing
   o Avoid graphics, columns, or dense blocks of text—keep it ATS-friendly

Use the "Bullet Plus" Formula

Don't just say what you did. Show how you did it—and what changed as a result.

Formula: What you did + How you did it + Result it created
Example: "Built GenAI-powered document summarization tool (what) using LangChain + GPT-4 (how), improving analyst productivity by 25% (result)."

This structure works across industries—and helps hiring managers quickly assess your impact.

Note: The "Bullet Plus" format—along with the action verbs in the next section—is a practical resume technique I first learned during my time at the University of Michigan Ross School of Business. It's part of the University of Michigan Career Center's public resume guidance and remains one of the clearest ways to frame impact for recruiters and hiring managers.

Power Up with Strong Action Verbs

The verbs you choose matter. They signal leadership, bias for action, and technical fluency. Instead of: "Worked on an AI feature" Try: "Shipped retrieval-augmented GenAI search capability across 5M+ queries/month" To help you craft impactful bullets, Figure 11.2: Action Verbs to Power Up Your GenAI PM Resume provides a range of categories and sample verbs to elevate your resume.

Here are a few categories to get you started:

Category	Action Verbs
Leadership	Led, Directed, Managed, Supervised, Administered, Coordinated, Organized, Facilitated, Oversaw, Guided, Orchestrated
Problem-Solving	Resolved, Solved, Improved, Enhanced, Innovated, Simplified, Streamlined, Corrected, Overcame, Transformed, Modified, Adjusted
Technical Skills	Adapted, Designed, Developed, Engineered, Fabricated, Programmed, Automated, Built, Created, Configured, Constructed, Implemented, Installed, Maintained, Debugged, Overhauled, Standardized, Upgraded
Communication	Presented, Negotiated, Persuaded, Advised, Educated, Formulated, Informed, Influenced, Consulted, Collaborated, Drafted, Authored, Published, Reported, Translated, Addressed, Solicited, Synthesized
Achievements	Achieved, Completed, Expanded, Exceeded, Improved, Pioneered, Spearheaded, Transformed
Helping	Adapted, Advocated, Aided, Assisted, Coached, Collaborated, Demonstrated, Diagnosed, Educated, Encouraged, Expedited, Facilitated, Guided, Helped, Supported, Volunteered
Analysis	Analyzed, Evaluated, Assessed, Interpreted, Calculated, Investigated, Measured, Audited, Examined, Forecasted, Surveyed, Studied, Tracked, Modeled
Project Management	Planned, Initiated, Launched, Executed, Delivered, Scoped, Coordinated, Prioritized, Scheduled, Measured, Controlled, Allocated
Creativity	Designed, Conceptualized, Innovated, Imagined, Visualized, Brainstormed, Formulated, Revamped, Revitalized, Pioneered
Research	Researched, Investigated, Identified, Discovered, Collected, Analyzed, Experimented, Surveyed, Interviewed, Tested
Operations	Operated, Managed, Controlled, Executed, Produced, Handled, Processed, Achieved, Accomplished, Completed, Delivered
Finance	Budgeted, Forecasted, Audited, Allocated, Managed, Projected, Reconciled, Appraised, Reduced, Expanded, Maximized
Customer Service	Assisted, Supported, Resolved, Consulted, Helped, Educated, Responded, Facilitated, Delivered, Addressed

Figure 11.2: Action Verbs to Power Up Your GenAI PM Resume

Let AI Help You Tailor—Intelligently

AI tools can speed up résumé customization—without losing your voice.

Here's how:

Extract keywords with ChatGPT or Gemini

Copy the job description and prompt: "Extract the top 10 keywords and required skills from this GenAI PM job description."

Now check: Are those terms reflected in your résumé?

Assess fit and gaps

Paste your résumé and the job description into ChatGPT and ask: "What gaps exist between this résumé and this job post? Which are critical vs. nice-to-have?" This helps you prioritize—and ensures you're not missing low-hanging credibility boosters.

Improve bullet points

Use prompts like: "Rewrite this bullet point to sound more impactful and highlight product thinking: 'Worked on LLM feature for customer service chatbot.'"

It might return: "Designed and launched LLM-based customer support flow using prompt chaining, reducing average response time by 35%."

Generate a job-aligned summary

Not sure what to write in the top summary section? Try: "Write a résumé summary for a GenAI PM applying to this job post based on this background." Then tweak it to reflect your real voice.

Maximize ATS compatibility

Applicant Tracking Systems (ATS) are used across companies of all sizes—from fast-moving startups to enterprises like Microsoft—to filter résumés before they ever reach a human. To increase your chances of getting noticed, use tools like Jobscan, Rezi, Teal, or SkillSyncer to align your résumé with specific job descriptions. These platforms highlight missing keywords (e.g., "RAG," "product management"), formatting issues, and compatibility risks. Aiming for a match rate of 75% or higher can significantly boost your odds of clearing ATS filters and moving forward in the hiring process.

From algorithms to humans

Clearing the ATS gate is only the first hurdle; your résumé still needs to resonate with the people who make the final call. To understand how recruiters and hiring managers evaluate candidates once the résumé is in their hands, see the quick primer below from *Jeremy Schifeling*—former recruiter/hiring manager at LinkedIn and Khan Academy, and author of Career Coach GPT. (See details in Appendix E—Item 6, "Inside the Recruiter 'Algorithm'")

1. Recruiters are risk-averse. They default to the safest, most obvious candidates.
2. De-risk yourself. Mirror the job's language, quantify results, and showcase proven impact.
3. Become an insider. Leverage alumni ties, warm intros, and networking to add human context to your résumé.

Final Tip: Keep a Tailoring Log

If you're applying to multiple roles, create a simple spreadsheet or Notion doc:

- Company name
- Link to JD
- Version of résumé you used
- Keywords included
- Proof point(s) shared

That way, you can track what works—and iterate fast. By combining strong fundamentals with modern AI tools, you'll create a résumé that doesn't just pass filters—it makes people want to talk to you. Let your résumé say: "This person isn't just interested in GenAI. They're already contributing."

You've Got Their Attention. Now Earn the Conversation.

A well-crafted résumé doesn't get you the job. But it earns you a shot—a moment of attention, a real conversation, a foot in the door. That's all you need. You've now got the structure, language, and tools to tailor your résumé with clarity and credibility—especially for roles where GenAI isn't just a buzzword, but a product reality.

Now the focus shifts.

Once you land that interview, everything moves from potential to performance. It's no longer about what you've written—it's about how you think, how you communicate, and how you show up in the room.

Let's get you ready for that.

Chapter 12: How to Prepare for GenAI PM Interviews

Ask: How do I turn potential into performance?

You've done the hard part—you earned the interview.

Now it's no longer about what you've written. It's about how you think, how you communicate, and how you show up in the room. This is where curiosity becomes credibility. Where vague interest becomes clear value. Where all the reflection and prep you've done so far gets tested—live.

But preparation isn't about cramming. It's about learning how to navigate ambiguity with structure. How to speak confidently about what you know—and honestly about what you're still learning. It's about walking into that conversation not trying to impress, but to connect.

This chapter will help you navigate the full journey—from managing interview logistics and researching the company to structuring your thinking, shaping your story, and building calm confidence under pressure.

You'll learn how GenAI PM interviews are evolving—from the kinds of loops companies run, to the signals they're looking for. We'll cover how to break down AI product scenarios without sounding robotic, how to resonate with both technical and non-technical interviewers, how to prep without over-rehearsing, and how to speak clearly about tradeoffs, risks, and unknowns.

Because GenAI hiring doesn't reward perfection. It rewards clarity, judgment, and the ability to think out loud—like a product manager.

How GenAI PM Interviews Are Evolving

Generative AI is not just changing the types of products we build—it's changing the way PMs are evaluated.

While many aspects of PM interviews still look familiar—product sense, execution, customer obsession—GenAI roles introduce a new layer of complexity. These interviews aren't just about shipping features in a predictable environment. They're about navigating unpredictability with confidence.

You're expected to reason through probabilistic systems, not deterministic outcomes. You'll be asked how you define success when outputs vary by the prompt, the model, and the context. You may be given a prompt like, "How would you design a GenAI experience for children?" and expected to discuss not just product-market fit, but also safety, explainability, and model guardrails.

Interviewers will probe your understanding of:

- Non-binary evaluation (What does 'quality' mean in a generative product?),
- Model constraints (Do you know what token limits mean for UX?),
- Real-world volatility (How do you ship in an ecosystem where models, APIs, and costs change monthly?), and
- Ethical complexity (How do you handle hallucinations, misuse, or bias—before it becomes a headline?)
- AI prototyping (Can you build and iterate on a simple GenAI feature live with an AI tool?)

Even the pace of the space becomes part of the test. GenAI evolves weekly—and interviewers want to know: how do you keep up? Are you just watching demos, or are you getting your hands dirty with prompts, APIs, or prototypes?

The Anatomy of a GenAI PM Interview Loop

GenAI Product Manager interviews still follow a familiar multi-round format—recruiter screen, technical deep dive, product case, cross-functional loop—but what's being evaluated at each step is shifting. In this space, titles are vague, tools evolve fast, and even the interviewers may still be figuring out the ideal profile.

But there's a pattern.

Whether you're applying to a scrappy AI startup or a scaled tech giant, most GenAI PM loops follow a multi-stage rhythm: early screens to assess clarity and intent, mid-loop case studies or design prompts to test structured thinking, and deeper dives on technical fluency, execution, and product judgment.

You're not just being evaluated on what you've done—you're being evaluated on how you reason when the ground shifts.

So what exactly happens at each step?

Let's walk through the typical GenAI PM interview loop—round by round—to understand what to expect, what's shifting, and how to approach each stage with clarity and confidence.

Recruiter or Initial Screen

This is where the conversation begins—but don't mistake it for a formality. The recruiter screen is a fit filter, and it matters more than most candidates think.

You're not being assessed for technical depth or deep product thinking just yet. You're being evaluated on story clarity, communication, and motivation. Can you explain why you're here? Can you articulate your value proposition in a way that feels focused, relevant, and sincere?

The recruiter is often the first advocate you need to win over. Their job is to assess whether you're worth pushing forward in a crowded funnel. If they can't confidently pitch you to the hiring manager afterward, the loop usually stops here.

You're not being tested on what you know—you're being tested on how you think when the ground shifts. That mindset starts here.

Show that you're adaptable, engaged, and aware of how fast GenAI is evolving—even if you're early in your own learning journey.

Common questions you will face

Here are some frequently asked recruiter screen prompts:

- "Tell me about your background—and why GenAI now?"
- "What are you looking for in your next role?"
- "Have you worked with LLMs, prompt engineering, or model-based systems?"
- "What kinds of AI products or companies excite you most right now?"

You don't need to have shipped a GenAI product to pass this screen. But you do need to show signal: intellectual curiosity, structured thinking, and the ability to connect your prior experience to what's needed in a GenAI PM role.

Answer with the PAST–PRESENT–FUTURE Framework

Use this simple but powerful story arc to structure your pitch:

PAST: What have you done so far that makes you credible?

- Highlight relevant PM experience, industry domain knowledge, or technical fluency.
- Example: "I spent the last three years leading NLP-based features in a B2B SaaS product."

PRESENT: Why this moment? Why GenAI?

- Show you're not just hopping on a trend. Tie your motivation to a clear insight or interest.
- Example: "After experimenting with GPT-4 and shipping a few side prototypes, I realized the future of product work is increasingly prompt-native and context-driven."

FUTURE: What are you looking for next?

- Be specific. Describe the type of team, challenge, or impact you're seeking—not just a job title.
- Example: "I'm excited to work at a company where I can contribute to shipping GenAI features responsibly at scale, ideally in a cross-functional team that blends research, infra, and UX."

Final tip: Prepare for curiosity, not perfection

You're not expected to know everything. You *are* expected to show you're learning quickly, thinking critically, and genuinely interested in building AI products that matter. Think of this round not as a test of mastery—but as a test of momentum. Show that you're not standing still.

GenAI PM Phone Screen

This is your first real product conversation—and a chance to move from "promising profile" to "high signal." Typically 45–60 minutes, this call is usually with a PM on the hiring team or a cross-functional partner like a data scientist. It's designed to assess your product thinking, communication clarity, and ability to work through messy, ambiguous problems with structure.

You're not being tested for deep technical knowledge yet—that comes later. This is about how you think: can you break down user needs, generate thoughtful product ideas, and reason through tradeoffs in a GenAI context?

You're not expected to have all the answers. But you are expected to ask the right questions—and think like a builder.

What interviewers are evaluating

- Can you translate vague prompts into structured thinking?
- Do you keep the user at the center—even when GenAI is in the room?
- Are you clear and intentional in how you explore tradeoffs and edge cases?
- Do you understand what GenAI enables, and how to wield that responsibly?

Common questions and prompts

This round tends to surface a mix of product sense and execution questions, often tailored to the company's domain. Some examples:

Product Design

- "Design a GenAI feature for Duolingo / Spotify / Uber."
- "Which GenAI product has impressed you—and why?"
- "How do you evaluate if a GenAI feature is actually valuable?"
- "If you were to reinvent search for GenZ using GenAI, how would you start?"
- "Walk me through how you'd build an AI-powered writing coach for students."

Execution & tradeoffs

- "How would you launch a summarization feature with limited model budget?"
- "You notice users are editing the AI output 50% of the time—what next?"
- "What would you prioritize if hallucinations increased after a model update?"

How to structure your thinking

You don't need a rigid framework, but structure matters—especially when answering product design prompts. A simple, flexible structure you can use in live interviews is:

Context > User > Pain Point / Problem > AI Opportunity > MVP > Risks > Metrics

Start by orienting yourself: What's the setting? Who are we solving for? What's broken? Then explore how GenAI could meaningfully help—not just by sprinkling it in, but by unlocking something that wasn't possible before.

For example, if you're asked to design an AI feature for Airbnb hosts:

"Let's anchor in the user: a host managing multiple listings and struggling to write personalized responses. GenAI could automate message drafting—using guest profile, tone, and stay context. The MVP could start with post-booking messages only, with a fallback to manual editing. I'd want to measure edit rate, response time reduction, and user satisfaction, while proactively monitoring for off-brand or biased outputs…"

This shows the interviewer not just that you understand GenAI—but that you know when to apply it, how to scope it, and what tradeoffs to watch for.

What makes a great answer stand out

You don't dive into AI tech first—you stay anchored in user needs. You make it feel like a product, not just a feature. You show awareness of what could go wrong, not just what could go right. You speak in a way that makes an engineer or designer say: "I'd love to build that with you."

You're not being tested on memorized answers—you're being tested on how well you can think on your feet, when the problem is fuzzy and the tech is still evolving.

Final tip: Think out loud, stay curious

If you hit a wall, say what you're thinking. Show your reasoning. Ask clarifying questions. You're not expected to be perfect—you're expected to be collaborative, structured, and curious. The best answers don't impress. They resonate. They show that you care about the problem, not just the tech. That you've explored the tools, but you're not beholden to them. That you're not just product-minded—you're problem-obsessed.

Case Study / Homework Assignment

Not every company includes this step—but when they do, it's often where the strongest candidates stand out. Case studies or take-home assignments are used to assess your depth, structure, and product leadership outside the pressure of a live interview. In some loops (e.g. Scale AI, Anthropic, Hugging Face), it replaces or complements one of the onsite rounds. In others—particularly at startups—it might be your biggest opportunity to demonstrate end-to-end thinking.

You might be asked to write a short product spec, respond to a prompted scenario, or deliver a slide deck walking through a new GenAI feature. The goal isn't perfection—it's to show how you think when you have time to reflect, prioritize, and structure. Think of this as a mini-product: not just what you build, but how you explain, scope, and defend it.

What interviewers are looking for

What stands out isn't a flashy design or a perfectly formatted deck—it's structured, user-centered thinking. Interviewers want to see that you're not just listing AI features, but making intentional, well-reasoned decisions. They're looking for clarity in what you prioritize, awareness of tradeoffs like speed vs. quality, and signals that you can lead a cross-functional team through ambiguity—not just follow a spec. GenAI should feel integrated into the solution, not layered on as decoration.

Case study formats and prompts

Product Design Brief

You might be asked to design a GenAI feature that helps customer support teams respond faster and better. A common format is a 1–2 page product brief or mini-PRD. It often includes a clear problem framing, user personas and flows, MVP scope, success metrics, and a fallback plan if generation quality isn't reliable.

Strategic Scenario

In other cases, you may be given a scenario like: *"You're a PM at a GenAI startup. OpenAI just launched a competing feature. What do you do?"* Here, the expectation is that you can reason through the market dynamics, adjust your roadmap, articulate a differentiation strategy, and briefly outline how your go-to-market approach would change in response.

Some companies ask for a short pitch deck—often around 10–15 slides. A typical prompt might be: *"Pitch a GenAI product idea for students and outline your 90-day plan."* Your response should walk through the core problem, proposed solution, why GenAI is a good fit, and how you'd define and measure success during the initial rollout.

How to structure your response

When working on a take-home assignment, shift from possibilities to decisions. Interviewers are no longer testing whether you *can* think of GenAI use cases—they're testing whether you can make smart bets and explain them clearly.

Use the same structure, but emphasize AI Leverage over AI Opportunity:

Context > User > Pain Point / Problem > AI Leverage > MVP > Risks > Metrics

Start by framing the product space and user. Define the pain point sharply. Then describe exactly how you'll leverage GenAI in your solution—what model or approach, why it's appropriate, and how it fits into the user flow.

For example: For customer support agents overwhelmed by message volume, we'll leverage GPT-4's summarization capability via Azure OpenAI to auto-generate message summaries and suggested replies. The MVP will focus on post-interaction wrap-ups. Risks include response accuracy and model misalignment with tone. We'll monitor edit frequency, task completion time, and CSAT to track success.

This shows you can reason like a product owner—not just explore ideas, but prioritize, scope, and de-risk them.

Note:

- Don't write a research paper. Write like you're briefing a cross-functional team.
- Use simple visuals (flows, wireframes, prioritization matrices) only if helpful.
- If it's a slide deck, aim for clarity over flash—think whiteboard, not Canva.
- Be explicit about tradeoffs: "I considered X, but chose Y because…"

What great candidates do differently

Strong candidates don't just describe *what* they'd build—they explain *why*, *how*, and *what might go wrong*. They think beyond the UI layer to show how the model actually works in context. They address ethical risks (like hallucinations or bias) proactively rather than defensively. And they keep the work crisp: clean structure, not 30 cluttered slides. A strong submission doesn't just inform. It inspires confidence—it makes the reviewer think, *"I want this person leading a team."*

AI Prototyping Loop

Possible: this could be folded into your Case Study or Final On-Site Loop, depending on the company's format.

This live-prototyping stage (30–45 minutes) tests your ability to move from concept to proof-of-concept—on the spot—using whatever toolset the interviewer provides. That might be an IDE or notebook (code), a low-code/no-code dashboard, or simply a prompt-chaining interface.

The goal isn't polished production code, but a clear, iterative prototype—code or no-code—that demonstrates your product sense, prompt-design skill, and real-time decision-making.

What interviewers are evaluating

- *Problem framing*: Can you translate a product prompt into a clear set of steps—whether in code, pseudocode, flow diagrams, or prompt sequences?
- *Prompt design & MVP focus*: Do you scope the minimal feature set and craft the right prompts or functions to deliver it?
- *Rapid iteration & trade-off awareness*: Can you pivot quickly—tweaking model parameters, UX elements, or logic—while balancing latency, cost, and output quality?
- Communication & collaboration: Do you explain your choices and next steps clearly, inviting feedback?

Common prompts and exercises

- "Prototype a GenAI-powered FAQ generator for a help center—either by wiring up a no-code interface or writing a few lines of pseudocode."
- "Build a simple chat interface that recommends personalized content, then iterate to add a guardrail against harmful outputs."

- "Sketch (or code) a mini workflow that classifies and summarizes incoming support tickets."
- "Show how you'd chain prompts or API calls to fetch data, generate a response, and render it in a basic UI stub."

How to structure your approach

- *Clarify the prompt*: Ask: Who's the user? What's the primary goal?
- *Outline your MVP*: Sketch the core steps or components (in a bullet list, pseudocode, or on-screen canvas).
- *Prototype*: Write minimal code, build in a low-code tool, or chain prompts in your interface.
- *Iterate aloud*: Tweak parameters, refine flows, or adjust UI as constraints emerge.
- *Summarize*: Recap what you built, why you chose that scope, and potential next improvements.

What makes a great prototype stand out

- *User-first scope*: You anchor in user needs before any code or tools.
- *Lean MVP*: You focus on the highest-impact functionality in the time allotted.
- *Thoughtful guardrails*: You anticipate risks (bias, hallucinations) and build quick checks.
- *Clear narration*: You talk through each decision so interviewers see your structured thought process.

Final tip: Think out loud, iterate fast

Whether you're writing code, wiring up a no-code dashboard, or chaining prompts, share your reasoning continuously. The best prototypes aren't perfect—they're evidence of smart, structured experimentation under tight constraints.

Final Loop (Design, Metrics, Technical, Behavioral)

This is the deep end.

The final loop—sometimes called "onsite," even when it's virtual—typically includes 3 to 5 back-to-back interviews, each focused on a different signal: product design, execution and metrics, technical collaboration, and behavioral or leadership judgment.

This is where GenAI PM interviews start to diverge more clearly from traditional PM loops. You're no longer just solving for usability or prioritization—you're solving for probabilistic systems, shifting model behavior, ethical ambiguity, and emerging UX patterns. And you're expected to do all that while still being... a great PM. *This isn't about knowing everything. It's about showing you can lead when no one does.*

Not all companies split these rounds cleanly, but the underlying signals remain consistent. Figure 12.1 outlines the most common structure and what each round is designed to evaluate.

Round	Focus Area	What They're Evaluating
Product Design	Ideation & structure	Can you frame user problems and apply GenAI responsibly?
Metrics & Execution	Product health, iteration, KPIs	Do you make smart tradeoffs? Do you think like an operator?
Technical Collaboration	Model usage, constraints, tradeoffs	Can you lead in partnership with AI/engineering teams?
Behavioral / Leadership	Teamwork, ownership, resilience	Are you thoughtful, accountable, and values-aligned?

Figure 12.1: Typical Final Loop Format

Sidebar: How the final loop varies across the GenAI stack

The structure of the final loop—as outlined in Figure 12.1—is broadly consistent across all GenAI PM roles. Whether you're working on infrastructure, models, applications, orchestration, or observability, you'll likely face rounds that test product design, execution, technical collaboration, and leadership.

What changes is what gets tested inside each round as outlined in Figure 12.2. A PM focused on inference infrastructure might be asked about latency metrics and GPU cost tradeoffs, while a PM building an agent orchestration layer may be asked about tool integration and fallback logic. Application PMs will spend more time on user flows and prompt UX, while model PMs might dive deep into fine-tuning, evaluation strategies, or safety constraints.

GenAI Stack Layer	Design Round Focus	Metrics & Execution Focus	Technical Round Focus	Behavioral Round Focus
Infrastructure	Usability of APIs, capacity mgmt, quota control	Latency, uptime, cost-per-token, service reliability	Hosting infra, GPU allocation, scaling bottlenecks	Navigating cross-team demands, ops vs. product tradeoffs
Model Layer	Use case fit, eval dataset design, customization	Hallucination rate, model accuracy, benchmark performance	Fine-tuning vs. RAG, prompt strategy, eval pipelines	Partnering with researchers, handling ethics/safety escalations
Application Layer	Workflow fit, UX prompts, feature scoping	Engagement, edit rate, CSAT, retention	Prompt design, token limits, user-facing model failure modes	Cross-functional product delivery, user advocacy
Orchestration Layer	Task decomposition, agent flows, tool chaining	Tool success rate, completion time, cost tracking	Agent memory, retries, fallback logic	Leading in high-ambiguity, coordinating across infra & apps
Observability/Governance	Dashboard design, access control, role-based UX	Coverage of evals, alert volume, bias or safety metrics	Red-teaming, model telemetry, logging strategies	Working with legal, risk, or compliance stakeholders

Figure 12.2: Interview Emphasis by GenAI Stack Layer.
© 2025 by the author of *Zero to GenAI Product Leader*.

The rounds stay the same. The questions shift based on where you sit in the stack. But the core signals remain: clarity, curiosity, structure, and sound judgment.

Product design: AI-first, user-driven

You'll likely be asked to design a GenAI feature or product. Your job isn't just to throw in AI—it's to ground the solution in user context, show disciplined reasoning, and account for what could go wrong.

Use this structured walk-through: *Context > User > Problem > AI Opportunity > MVP > Risks > Metrics*

Example prompt: "Design a GenAI feature to help job seekers improve their resumes." You might say:

"Job seekers often struggle to tailor resumes for each role. The opportunity here is to use an LLM to generate custom summaries and bullet points based on the job description. We'd start by limiting scope to one job type (e.g., tech), one format, and one tone. MVP metrics would include usage, edit rate, and downstream interview callbacks. Risks include hallucination, formatting bugs, and bias in output—which we'd mitigate by letting users review and edit before submission."

This round isn't about dazzling the interviewer. It's about showing you can navigate ambiguity while staying grounded in user value and execution feasibility.

Metrics & execution: From intuition to insight

Here, you'll face questions about how you measure success, manage tradeoffs, and iterate. With GenAI products, this isn't straightforward—success might mean low hallucination rates, high completion without edits, or even user trust over time.

Common prompts:

- "What metrics would you track for a GenAI writing assistant?"
- "Your summarization feature shows high engagement but poor satisfaction. What now?"
- "If instrumentation goes down mid-launch, how do you assess success?"

Show that you:

- Understand leading vs. lagging indicators
- Can work with proxy metrics if direct signals aren't available
- Have a point of view on experimentation in AI systems (e.g., A/B tests, human feedback, offline evaluation)

Use the unified framework again—especially the Risks > Metrics section—to drive the conversation toward practical, measurable outcomes.

Technical + domain fluency round

Note: This isn't a coding interview—but it isn't a casual chat, either. You won't be whiteboarding algorithms, but you will be expected to think and speak like a product builder in GenAI.

What's being tested here is your ability to reason like a product builder in the GenAI space. That means understanding how the system works, where it breaks, and how product decisions shape outcomes—even if you didn't write the code yourself. Many companies now fold live prototyping (see *AI Prototyping Loop* earlier in this chapter) into this round, so be prepared for both discussion and demo.

You're not expected to be a deep technical expert. But you are expected to:

- *Think at a systems level:* understand how components fit, where they fail, and how product choices shape outcomes.
- *Collaborate with engineers:* frame trade-offs clearly and lead technically informed decisions.
- *Demonstrate AI/product fluency:* Craft effective prompts (temperature, max_tokens, role messages), scope lean MVPs via low-code or prompt-chaining tools (e.g., bolt.new, Lovable), define AI-specific metrics (perplexity vs. relevance, token efficiency, hallucination rate), and build ethical guardrails (bias filters, moderation, human review).

Some roles do require deeper fluency (infrastructure, model dev, evaluation pipelines). In those cases, you may be asked to reason about GPU allocation, latency optimization, benchmark design, or model configuration workflows. You don't need to code, but you do need to speak the language of the system—and lead confidently in technical conversations.

Importantly, interviewers may not quiz you directly on "What's a vector database?" or "How does RAG work?" Instead, they'll say things like:

- "Tell me about a time you worked on a technically complex product."
- "How did you partner with data science or engineering on feature design?"
- "What tradeoffs did you face when balancing performance, privacy, or cost?"

This is your moment to connect product thinking with technical fluency—even if your experience came from side projects, hackathons, or self-study.

Maybe you:

- *DIY Prototypes:* Built a mini demo (e.g., GPT + Zapier, bolt.new, or Lovable flow) to explore prompt chaining and basic guardrails.
- *Analytical Teardowns:* Created a teardown analyzing hallucination risks in a public LLM tool

- *Course & Project Work:* Completed a GenAI course project (mock PRD, fine-tuning exercise) and translated insights into product recommendations.
- *Agent Experiments:* Assembled a simple agent using LangChain, Semantic Kernel, or a low-code stack (Flowise, CrewAI + Replit) to automate a multi-step task (e.g., pulling data from a doc, summarizing it, formatting a report, and auto-sending it via email or Slack)
- *System Deep-Dives:* Simply spent time breaking down GenAI architectures in side projects or portfolio posts

That's all fair game—if you can speak credibly about what you learned. You're not here to recite jargon. You're here to show you understand how the system behaves—and how that behavior affects UX, safety, performance, and product strategy.

If you don't have GenAI-specific experience yet, lean on adjacent proof:

- Times you worked with engineers or scientists in ambiguous domains
- How you evaluated tradeoffs in data-heavy or privacy-sensitive products
- Stories where you had to learn fast, ask smart questions, and build trust with technical teammates

Remember: the goal here isn't perfection. It's pattern recognition.

- Can you explain system behavior in plain English?
- Can you explain why a model might hallucinate—and how that affects user trust?
- Can you distinguish between experiments and production-grade solutions?

That's what technical fluency looks like in GenAI PM interviews. It's not about memorization. It's about showing that you think like a builder—and that you're not afraid to get close to the system.

Behavioral & leadership: PM judgment in the wild

Finally, you'll face questions that probe how you lead—especially under pressure, with ambiguity, and across disciplines. These are often situation-based:

- "Tell me about a time you shipped something you weren't confident in."

- "Describe a disagreement with an engineer or researcher and how you navigated it."
- "How do you make tradeoffs between innovation and safety?"

Use the STAR method but go beyond just listing what happened. Show what you learned. Show how you've grown. And if your experience is limited in GenAI, show how quickly you've leveled up and contributed meaningfully anyway. This round often decides whether they see you as a safe bet—or a strategic multiplier.

Final tip: Show judgment, Not just fluency

You don't need to be the most technical person in the room. But you do need to show that you can lead in a room full of uncertainty—and make decisions that balance user value, technical feasibility, business priorities, and ethical responsibility. *You're not expected to have all the answers. You're expected to have a compass.*

Leadership panel (Optional)

Not every loop includes this step—but when it does, it's usually the last and highest-leverage conversation. You're no longer being evaluated just on skills. You're being evaluated on judgment, maturity, and alignment with the company's direction.

This round often includes one or more senior leaders—a Director of Product, Head of AI, or even the CEO or CTO at a startup. They're not looking for a walkthrough of your resume. They're testing how you think at altitude, how you prioritize under pressure, and how you'll help the company win—especially in a rapidly evolving GenAI space.

What they're looking for

- Can you zoom out from feature-thinking to strategic tradeoffs?
- Do you understand the broader GenAI landscape—competitive dynamics, business models, emerging risks?
- Can you communicate clearly and credibly across functions—including to non-technical execs?
- Do you show sound product judgment, not just tactical execution?
- Are you someone who will thrive in *this* organization—not just any PM role?

This round is often the most conversational. But that doesn't mean casual. It means you need to come in prepared to talk strategy, ready to own your story, and able to engage as a peer, not just a candidate.

Common questions

You'll often face a blend of strategic and reflective questions. For example:

- "Where do you think GenAI is headed in the next 3–5 years?"
- "What's a GenAI product or company you admire—and why?"
- "How do you decide what not to build?"
- "Tell me about a time you influenced a difficult stakeholder."
- "If you joined tomorrow, what's the first thing you'd want to fix or improve?"

You might also be asked questions like:

- "How do you think about safety and ethics in GenAI products?"
- "What are the biggest risks you see in how companies are adopting GenAI right now?"

How to show up strong

- Speak in systems, not slogans. Don't say "I'd prioritize user trust." Say, "I'd measure user edit rate and retention after AI suggestions to proxy trust, and I'd design opt-outs or explainability where appropriate."
- Tie ideas to outcomes. If you're pitching a direction, anchor it in metrics, GTM motion, or user behavior—not just "cool tech."
- Connect your story to the moment. Why GenAI? Why now? Why this company? This is your chance to signal both momentum and intentionality.

This round is often where final doubts get resolved—or new doubts are introduced. You're not just here to prove that you can do the job. You're here to show that you'll grow with the company—and help it navigate a space that's still being defined.

Real-world Interview Breakdowns

Most GenAI PM loops follow a familiar rhythm—recruiter screen, phone screen, case or take-home, and a final loop—but how each company runs it depends on what they value most. Let's look at how four organizations—OpenAI, Scale AI, early-stage GenAI startups, and Big Tech companies—approach interviews in practice. Each reflects a different slice of the GenAI ecosystem: foundational model builders, fast-moving product innovators, scaled platform players—and companies like Scale AI that power the infrastructure layer in between.

While not as publicly visible as OpenAI or as globally scaled as Microsoft or Google, Scale AI plays a foundational role in the GenAI value chain—powering model evaluation, data pipelines, and labeling workflows that sit at the operational heart of many leading AI products. For candidates interested in applied GenAI systems—where tooling, optimization, and real-world constraints collide—Scale's interview process offers a valuable window into the kinds of product decisions that happen behind the curtain.

OpenAI

For mid-to-senior PM roles, OpenAI runs a structured five-round process: recruiter screen, PM phone screen, a take-home assignment, and a multi-part onsite loop. The focus is on your ability to think clearly in high-ambiguity environments, balance user value with safety, and partner effectively with deeply technical teams.

The onsite often includes:

- A product design round ("Design an AI feature for education that scales globally"),
- A product execution deep dive (e.g., managing a roadmap across multiple models),
- A technical fluency round (e.g., tradeoffs between RAG and fine-tuning),
- A risk/ethics discussion (e.g., handling unsafe output in high visibility launches),
- And a leadership panel, where you're asked to share how you'd shape the future of GenAI product development.

Candidates who do well here tend to demonstrate clarity, grounded optimism, and comfort navigating ambiguous tradeoffs—especially where technical and ethical dimensions collide.

Scale AI

Scale's loop for APM and mid-level PM roles emphasizes systems thinking, internal stakeholder management, and go-to-market clarity. The structure is usually four rounds: recruiter screen, PM phone screen, a take-home spec or deck, and a final loop that blends design, execution, and collaboration.

You might be asked to:

- Design a GenAI feature for enterprise users (e.g., improving annotation flows),
- Build a short take-home (slides or spec) for a labeling tool improvement,
- Whiteboard tradeoffs between model accuracy and labeling efficiency,
- And discuss alignment with stakeholders—including engineers, ops teams, and even the CEO.

Strong candidates demonstrate they can balance speed, safety, and scale—not just ship features.

Early-stage GenAI startups

Startups tend to move fast and expect candidates to show ownership and product creativity from day one. Interview loops are leaner—often just 2–4 steps—and may involve direct interaction with founders.

Expect a lightly structured phone screen (often with the founder) asking what you've built or explored recently, followed by a live case exercise ("Sketch a GenAI agent that helps freelancers manage workflows"), and possibly a short take-home (a Notion doc, Figma prototype, or slide deck). The last step is often a culture-fit conversation, where you'll be asked what you'd prioritize in your first 30 days. In these interviews, side projects often speak louder than résumés. If you've built even a simple tool—like a GPT-powered assistant or teardown of an existing GenAI product—bring it up. Startups want to see bias toward action and signs of scrappy, systems-aware thinking.

Big tech (e.g Microsoft, Google, Meta etc)

For GenAI PM roles at large tech companies, the interview loop follows a structured, multi-signal format—often with 5–6 rounds covering product design, execution, metrics, technical collaboration, and cross-functional leadership. These companies are building GenAI products at global scale, often with significant internal complexity, regulatory oversight, and long-term platform implications.

But don't let the structure fool you—ambiguity is everywhere.

Even in big tech, the landscape is shifting constantly. New models are released, APIs evolve, policies change, and user expectations remain fluid. The strongest candidates show they can operate with clarity inside that ambiguity—not by waiting for direction, but by proposing direction and navigating tradeoffs with confidence.

A note on Amazon

Amazon is a unique case among big tech companies. All PM interviews—including GenAI roles—are behavioral by design, deeply rooted in the company's 16 Leadership Principles. Unlike other companies that might alternate between case-style and resume-driven rounds, Amazon interviews focus almost entirely on past experiences. You'll be expected to show depth, specificity, and clear outcomes across questions like:

- "Tell me about a time you challenged a senior stakeholder and were right."
- "Give an example of how you took ownership and delivered under tight constraints."

Each answer is assessed not just for outcome, but for the right level of detail—demonstrating scope, influence, judgment, and consistency with Amazon's bar-raising expectations. In GenAI roles, expect those same behavioral standards to apply—but adapted to scenarios involving uncertainty, rapid iteration, risk tradeoffs, or responsible AI decisions.

Interview Prep Framework: From Insight to Action

Before you dive into whiteboards, mock interviews, or frameworks, take a moment to orchestrate the process. The logistics around scheduling your interviews—often overlooked—can shape how well you perform. If you're interviewing at a company like Amazon, where the bar is high and every round is behavioral, pacing and preparation matter even more.

You're not just being evaluated on your answers—you're being evaluated on your energy, clarity, and judgment under pressure. That starts before the first question is asked.

Scheduling your interview strategically

When Company X reaches out, your instinct might be to say "I'm ready anytime." But smart candidates create space to prepare. Here's how to approach it:

Time zone awareness: If your interviewers are spread across time zones, suggest slots that align with your peak alertness—not just what's available.

Prep buffer: Don't hesitate to ask for 3–5 days to prepare, especially for case studies or multi-round loops. Framing it as "I want to do this thoughtfully" often earns respect.

Multi-round navigation: If the loop includes back-to-back sessions (e.g., product design, technical, leadership), see if they can be spread across two days. This helps you show up fresh, not fatigued.

Communicate early and clearly: HR teams appreciate candidates who confirm logistics, ask who's on the panel, and clarify what's expected (e.g., live whiteboarding, or a GenAI case).

Once the logistics are set, you shift to real work

You've aligned the schedule, set expectations, and carved out prep time. Now, it's time to get tactical about how you'll prepare—not just vaguely "study" for the interview but structure your prep with intent. That's where the PREPARE framework comes in.

Introducing the PREPARE framework

To guide your entire GenAI PM interview prep, this chapter introduces the PREPARE framework—a practical sequence to help you move from confusion to confidence:

- P – Product and Company Research
- R – Review Generative AI Concepts
- E – Evaluate & Align Your Experience
- P – Prepare Stories for Behavioral Interviews
- A – Articulate Responses for Non-Behavioral Questions
- R – Role-Play and Mock Interviews
- E – Engage with Questions for the Interviewer

Each step will help you layer confidence on top of clarity, so that by the time you walk into the interview, you're not just hoping to perform—you're ready to lead the conversation.

P – Product and company research

Before you can perform well in a GenAI PM interview, you need to understand what problem the company is solving, where AI fits, and how your skills can move the needle.

This isn't about memorizing their mission statement. It's about going deeper— understanding how GenAI shows up in their product strategy, what kind of PM they likely need, and what tradeoffs they're navigating right now.

Start with their stack and strategy

Use this checklist to reverse-engineer what kind of GenAI PM the company might be looking for:

- What layer of the GenAI stack do they operate in? Are they building foundational models (e.g., Mistral, Cohere), orchestration platforms (e.g., LangChain, Semantic Kernel), vertical agents (e.g., Harvey for legal), or AI-enhanced features inside SaaS products?
- What user problems are they solving with GenAI? Are they focused on productivity, summarization, automation, multimodal search, customer support, or internal workflows?
- What form does AI take in their product? Is it embedded in the UI? Exposed via APIs? Used behind the scenes to power internal decision-making?
- Are they monetizing GenAI directly or using it as a differentiator? Are they charging for prompts, tokens, subscriptions — or using AI to drive engagement and retention?
- Do they publish updates about their GenAI direction? Check their product blog, engineering blog, press releases, GitHub, and conference talks (e.g., OpenAI Dev Day, Google I/O, Microsoft Build).

If the company is public, review their 10-K or annual report. Scan for how leadership discusses GenAI in their strategy, innovation, or risk sections. Are they investing heavily? Cautious? Do they view AI as core to their moat?

Understand the culture and context

To get a feel for how the company thinks about innovation, start by scanning their *About* page, product blog, or engineering updates. You'll often find hints about how seriously they treat AI—whether it's core to the roadmap or more of a brand play.

For cultural signals, platforms like Glassdoor, Levels.fyi, or Blind can help you gauge the company's pace, values, and leadership tone. And if they've recently raised funding, announced new AI partnerships, or hired leaders from major labs, that's usually a strong indicator that GenAI is a strategic priority—not just a feature layer.

Analyze the role: Go beyond the bullet points

Job descriptions often sound generic—until you start reading between the lines.

Pay attention to what kinds of decisions this PM will likely own. Are they shaping data pipelines? Defining LLM-driven features? Handling model evaluation workflows or internal tooling? These cues tell you whether the role is close to infra, model, orchestration, or application layers.

Look for cross-functional signals too. Mentions of collaboration with research, infra, or legal teams hint at the complexity of the org—and the judgment required to navigate it. If ethical AI, explainability, or governance appear in the description, you're probably looking at a role where compliance and risk are first-class product concerns.

Even the verbs matter. If the JD says "scale" or "optimize," you're stepping into an existing product. If it says "define," "explore," or "prototype," you may be the first PM on something new. Matching your mindset—0 to 1 builder or 1 to n operator—to their ask can give your interview answers more authenticity.

And if the company has published research, contributed to benchmarks, or open-sourced internal tooling, reference one in your interview. A single thoughtful mention can make you stand out.

Create a 1-pager to sharpen your thinking

This isn't something you need to carry into the interview—but creating it will help crystallize your research and strengthen your responses. Think of it as a personal briefing doc, not a handout.

Include:

- The company's current GenAI footprint
- Their likely product constraints or bets
- Your own experience (or projects) that align with those pain points or use cases
- One well-informed observation or question to ask during the interview

Keep it concise. The goal is to walk into the interview with sharp mental recall, not rehearse a monologue.

Focus your research around what matters

You don't need to become an expert in everything they've ever built. Instead, center your research around a few key questions:

What GenAI capabilities do their users actually value? Where are they ahead—or lagging behind competitors? What product or UX frictions might they be facing around cost, latency, or hallucination? Are they shipping features that showcase new model capabilities—or mostly wrapping existing ones?

Also check their recent AI partnerships, public benchmarks, or open-source contributions. These can signal how much they're investing in foundational vs. applied innovation—and give you ideas for how to position yourself as someone who can accelerate that work.

Researching Agentic AI Products?

If the company builds AI agents—tools that plan, act, and take decisions on a user's behalf— your research needs to go deeper than just outputs. Look for signs of orchestration: Are they using LangGraph, Semantic Kernel, or CrewAI? Are agents executing multi-step tasks, invoking APIs, or making autonomous choices? Try to reverse-engineer what kinds of user problems require reasoning, memory, or tool use—and how the company balances autonomy with control.

R – Review GenAI concepts

Before you can speak confidently about AI, you need to ground your understanding in the specific context of the role. Start by revisiting the core stages of any GenAI workflow—data ingestion, model selection or fine-tuning, prompt design, inference, and post-processing—with the company's product in mind.

Sketch a simple flow diagram showing how their users' data might move through each stage, and annotate it with the trade-offs they face (for example, latency versus accuracy in a real-time chat app, or cost versus customization in a fine-tuned analytics feature).

As you build your concept map and glossary, don't just outline data flows and model behaviors—also flag the business impact at each stage. For example, note whether a prompt-tuning shortcut could save X percent on inference cost, or if tighter drift monitoring would reduce support tickets by Y percent.

Next, build a one-page glossary of the terms most relevant to the job—say "RAG," "temperature," "vector database," or "hallucination"—and write a concise, role-focused definition for each. Rather than a dry dictionary entry, frame each term through the lens of the company's product: how would "temperature" settings affect their customer-facing summarization tool, or where does "hallucination" risk show up in their AI agent?

Then, practice explaining these concepts out loud as if you're in the interview room. Record a 60-second pitch for three of your glossary terms, using plain language and tying every example back to the company's domain. Finally, spend a short session in a no-code playground (bolt.new, Lovable, or OpenAI Playground) running a prototype prompt that mirrors the role's use case—observe how tweaking parameters shifts the output, and note those insights in your concept map.

By the end of this exercise, you'll have a portable GenAI prep kit—a diagram, a tailored glossary, and first-hand prototype learnings—that you can reference to demonstrate both depth of knowledge and role-specific relevance under pressure.

When outputs become actions?

In Agentic AI, you're not just generating responses—you're triggering behaviors. Brush up on how tools like ReAct, AutoGen, Semantic Kernel and LangChain structure planning, tool invocation, and step-by-step reasoning. Interviewers may expect you to reason through tradeoffs like: Should an agent take the next step automatically? What happens when it loops, fails, or misfires? Fluency here means understanding how AI output becomes real-world impact—and how to build safe, recoverable systems.

E – Evaluate & Align Your Experience

Map your best work to GenAI realities—so you arrive knowing exactly which wins to spotlight and how they translate to an AI-driven role.

You've already reflected on your strengths and mapped your fit (Chapter 10), and you've shaped your story to get in the door (Chapter 11). Now it's time to get interview-ready—to express those same instincts —problem framing, prioritization, execution, user empathy —in an AI-native context. You're not relearning PM fundamentals; you're reframing your proven skills to navigate probabilistic outputs, evolving models, and ambiguous success metrics.

In this step, you'll retrofit your top achievements, so interviewers immediately see how your wins translate to GenAI challenges. Think of E as the ideation and mapping phase: you're gathering the right raw material. In the next step (P), you'll polish those examples into full STAR stories.

To turn this idea into reality, complete these three actions:

Identify Your Top Experiences

Scan your resume for the handful of projects that best showcase your core instincts—problem framing, trade-off ownership, data-driven impact, ambiguity management, or cross-functional leadership. For each, note the one or two concrete aspects you'll want to highlight in discussion (e.g., "I reduced latency by 30%," "I led a multi-team rollout," "I designed a dashboard to track user satisfaction").

Build Your GenAI Lens with AI Help

Take your list of top experiences and the job description for your GenAI PM role. In ChatGPT, Groq, or another AI tool, provide both lists and ask:

"Here are my past projects [paste bullet-list] and the GenAI PM role I'm interviewing for [paste role summary]. Which transferable skills and accomplishments should I highlight to demonstrate fit, and how would each apply in an AI-driven context?"

Refine the tool's suggestions into a crisp, one-line "GenAI lens" annotation for each experience *(e.g., "I'd prototype prompt chains to balance faster response times against API cost—then monitor hallucination rate to ensure quality.")*

Shortlist for Deep Dive

Rank your GenAI-annotated examples by how closely they align with the role's focus (infrastructure, orchestration, or end-user applications). Pick the top two or three. Those become your raw material for the next step, **P – Prepare Behavioral Stories**, where you'll craft full STAR narratives—rooted in your real work but ending with that AI-powered closing insight.

By completing **E**, you'll finish with a set of ready-to-use "flashcards" pairing your proven impact with domain-specific AI trade-offs and metrics—your secret weapon for answering both "Tell me about a time…" and "How would you…" questions with instant GenAI fluency.

P – Prepare stories for behavioral interviews

Show how you lead when things aren't easy

Technical fluency might get you noticed—but your storytelling is what makes you memorable. Behavioral interviews are where hiring managers evaluate your leadership instincts, judgment under pressure, and ability to navigate messy situations—especially the kind that GenAI roles are full of.

If your answers sound rehearsed, vague, or too safe, you'll blend in. If your stories are honest, structured, and insight-rich, you'll stand out.

Behavioral questions aren't about your past—They're about your readiness

Contrary to popular belief, behavioral interviews aren't designed to test how much you've accomplished. They test how you operate: how you make decisions, manage friction, communicate uncertainty, and recover from mistakes.

In GenAI PM roles, this matters even more. You're working in uncharted territory—with fast-changing tech, unclear policies, and emergent risks. Your ability to navigate ambiguity isn't a bonus. It's the job.

Use the STAR+L method: Add what most candidates miss

You've likely seen STAR—Situation, Task, Action, Result—before. But in GenAI roles, there's one more element that separates good from great: Learning. What surprised you? What would you do differently now?

Use STAR+L as your structure:

- Situation: Set the context. What was happening?
- Task: What were you responsible for?
- Action: What did you do—and why that approach?
- Result: What changed? How did you know it worked?
- Learning: What did you take away? How did this shape how you lead now?

This last step shows humility, awareness, and growth—all signals that you can lead in a domain that evolves by the week.

Focus on the right kinds of stories

You don't need to have built GenAI products to tell relevant stories. You just need to show that your instincts map well to the work. Make sure you've prepared at least one solid story for each of these themes:

- Ambiguity navigation: A time when the problem wasn't well defined—and you helped shape the solution space.
- Cross-functional alignment: A moment when teams had conflicting views—and you helped converge in a direction.
- Failure recovery: When something broke post-launch, or user feedback challenged assumptions — and you led the response.
- Risk awareness: How you handled privacy, safety, compliance, or user trust in a high-stakes feature.
- Learning velocity: A situation where you ramped up quickly in a new domain or built credibility without being the expert.

Bonus if your stories include work with researchers, infra teams, designers, or legal/compliance—all common collaborators in GenAI product work.

Cultural fit and collaboration still matter

Behavioral interviews are also where teams assess how well you'll work with them—especially when you're not the expert in the room. Stories that highlight:

- How you resolve tension with engineers or researchers
- How you communicated with execs or designers when priorities diverged
- How you made tradeoffs without perfect data

...send a signal that you're collaborative, systems-aware, and not just "driving features"—but building relationships and trust.

Final thought: The best stories don't make you look perfect—They make you look prepared

GenAI PMs don't get rewarded for being the smartest in the room. They get trusted because they lead calmly through chaos. You'll face moments where a model drifts, a stakeholder pushes back, or a user's trust is on the line. The question your interviewers are asking is:

"Will this person step up when it matters?" Your stories are how you answer—without saying it directly.

A – Articulate responses for non-behavioral interview questions

Build and rehearse clear, adaptable structures for the "what" questions—product design, execution & metrics, strategy, and technical collaboration—so you can answer crisply under pressure.

While behavioral questions focus on your past, non-behavioral questions test how you think in the moment. These include product design, execution, metrics, strategy, and technical fluency—especially in GenAI-specific scenarios. Your interviewer isn't just checking your answer—they're evaluating **how** you break down ambiguity, organize your thinking, and arrive at a thoughtful, outcome-focused response.

This section will help you respond like a GenAI product manager: calmly, logically, and with judgment rooted in product outcomes—not buzzwords.

Build your question inventory

Before you can answer crisply, spend 20–30 minutes compiling real interview prompts from your Anatomy notes and the job description, from Glassdoor, Blind, Levels.fyi, or LinkedIn comments on GenAI PM interviews, and from peers' mock-interview debriefs or industry forums.

Once you've gathered a robust list, group them into four categories—product design (for example, "Design a GenAI résumé coach"), execution and metrics (for example, "How would you measure success for this AI feature?"), technical collaboration (for example, "How would you work with engineers on latency and drift?"), and scenario prompts (for example, "With only 10 API calls per minute, how do you optimize for quality?"). This inventory becomes the foundation of your targeted practice.

Choose your framework

Anchor every answer in this adaptable sequence—starting with a quick clarification and ending with metrics:

0. Clarify & paraphrase: If a prompt feels broad or ambiguous, pause to restate or ask a follow-up:
    a. "Just to confirm, are we prioritizing user satisfaction or cost efficiency?"
    b. "So you'd like me to focus on latency versus quality—correct?"

1. Context: Describe the business or user scenario
2. User: Identify who you're serving and their core need
3. Problem: Pinpoint what's broken or missing
4. AI Leverage: Explain how GenAI uniquely solves it
5. MVP: Outline the simplest first version
6. Risks: Call out potential failures and safeguards
7. Metrics: State how you'll measure success and catch issues

Treat these steps as guideposts, not a script.

The best PM candidates flex their frameworks like tools, not templates. For example, if you're asked: "Design a GenAI feature to help job seekers write better resumes," a strong response might naturally cover:

- **"Who** the target users are and what pain they face"
- **"How** GenAI can augment that workflow (without overpromising)"
- **"Key constraints** (e.g., hallucination, bias, tone, formatting)"
- "A **focused MVP** and rollout plan"
- **"Metrics** that track both value and risk"
- "How you'd handle **edge cases or feedback loops**"

Tailor the flow to each category

Not every question needs all seven steps. Lean into the slices that matter most:

- For **product design**, emphasize Context > User > AI Leverage > MVP.
- For **metrics and execution**, focus on Problem > Risks > Metrics > Course correction.
- For **technical collaboration**, highlight Problem > AI Leverage > Risks → Handoff/Alignment.
- For **scenario prompts**, drill into Context > MVP > Metrics > Next-step plan.

As you practice each, drop in one "AI trade-off" insight (your GenAI lens)—for example, balancing cost versus latency or accuracy versus hallucination risk.

Connect to Strategy

Every answer is stronger when tied back to the company's goals or KPIs. Weave in a business metric or strategic priority you uncovered in your research: "This

MVP not only reduces churn but also aligns with your Q4 goal to accelerate enterprise adoption."

Get comfortable with ambiguous metrics

Many GenAI features don't have obvious KPIs—especially when outcomes are subjective. You might be asked:

- "How would you measure success for a code generation tool?"
- "What if engagement is high but satisfaction is low—what next?"
- "How would you test whether AI suggestions are helpful or just perceived as helpful?"

In your response, show that you can:

- Differentiate between leading and lagging indicators
- Use proxy metrics when instrumentation is limited
- Mix quantitative and qualitative signals (e.g., edit rate, trust scores, feedback volume)
- Incorporate fallback plans when tracking fails

GenAI products often behave unpredictably—and so does the data. Great PMs can make forward progress without perfect signals.

Final thought: Don't just solve—prioritize and explain

In GenAI interviews, you're not being tested on whether you get the "right" answer. You're being tested on how you frame problems, weigh options, and communicate under uncertainty. Don't aim for the cleverest solution. Aim for the clearest one. Then explain why you chose it.

R – Role-play and mock interviews

Close the gap between knowing and showing

You've studied the questions. You've written your stories. You've mapped your frameworks. Now comes the step that turns knowledge into performance: practicing it live.

GenAI PM interviews are rarely predictable. You won't always get clearly scoped prompts. You might be asked to design a feature for hiring managers, only to be interrupted mid-answer with a twist: "What if the model starts hallucinating during onboarding?" Or you might be in the middle of laying out a strategy when your interviewer throws in a curveball: "Your competitor just launched the same thing—now what?"

These moments aren't designed to trip you up. They're designed to see how you think on your feet. Role-playing helps you build the fluency to respond without freezing, the structure to avoid rambling, and the presence to stay clear even under pressure.

The goal isn't to memorize answers—it's to build the muscle of speaking productively when the path isn't clear. It's your dress rehearsal for real conversations.

How to practice—and what to focus on

Start by alternating between behavioral and non-behavioral questions. One moment you might be asked to describe a time you handled a product failure; the next, you're expected to reason through an ambiguous GenAI feature request.

To simulate real conditions, set a timer and treat each question like the real thing. Time yourself: five to seven minutes per answer. Better yet, ask a peer, mentor, or even an AI tool to act as the interviewer. Ask them to interrupt you. Push back. Ask follow-up questions. The goal is to practice not just delivery—but adaptability.

Who to practice with

- Peers in product will give you useful feedback on clarity and flow.
- Mentors or coaches can help push you beyond surface polish into strategic depth.
- AI mock platforms (like Final Round AI or Exponent) are good for self-paced drills, especially if you're short on human practice time.

Even something as simple as recording your own response and playing it back will reveal whether your stories sound clear, grounded, and credible—or if they drift, meander, or feel overly rehearsed.

What to listen for in playback or feedback

- Are you rambling or clear?
- Are you overly mechanical, or do you sound authentic and engaged?
- Are you using structure—without sounding robotic?
- Are you tying technical concepts to user or business value?
- Are you ending each answer crisply, with a confident signal?

Every rep helps. Great PMs don't deliver perfect answers—they think aloud in structured, human ways. They lead the room even when the prompt is vague.

Final thought: Real interviews are conversations, not exams.

The best interviews don't feel rehearsed—they feel thoughtful, alive, and collaborative. You're not being scored on a script. You're being evaluated on whether the interviewer would want to work through hard problems with you. Mock interviews help you stop "presenting" and start thinking with people. So don't just prep. Play. Mess up. Start over. Practice like you ship products—iteratively.

E – Engage with questions for the interviewer

Signal curiosity, confidence, and strategic alignment

In a GenAI PM interview, the final minutes aren't filler—they're leverage. When the interviewer asks, *"What questions do you have for us?"*, it's not a throwaway moment. It's a signal-check. One last chance to show that you've done your homework, that you're thinking beyond the role, and that you're already engaging like a future collaborator. This moment becomes even more powerful in Agentic AI roles, where you're not just managing feature delivery—you're shaping system behavior. Your questions should reflect systems thinking, an understanding of ambiguity, and an ability to navigate tradeoffs at both the model and product layers.

What interviewers infer from your questions

The way you ask reflects the way you think. When you come prepared with thoughtful, role-specific questions, you're signaling:

- Technical and strategic fluency
- Curiosity about ambiguity, not just clarity
- A builder's mindset—someone who wants to co-own the future

And in roles involving AI agents, orchestration, or tool-use, your questions should show you're already thinking about: *How does the system decide? Where does it fail? How do users stay in control?*

Framework for creating high-signal questions

A reusable structure to craft strategic, company-relevant questions in any GenAI PM interview

You already know that the final minutes of an interview are a powerful signal-check—especially in GenAI roles where ambiguity, orchestration, and autonomy are part of the product surface. Now it's time to equip yourself with a durable framework you can apply in any setting—across different company sizes, interviewer types, and AI stacks.

Great questions follow a pattern. Not to sound smart—but to surface nuance, engage as a future teammate, and clarify the road ahead. Here's how to construct questions that do exactly that.

"High-Signal Questions = Context + Curiosity + Relevance" - This simple structure (see Figure 12.3) helps you construct questions that are anchored, thoughtful, and relevant to the role.

Component	What It Means	Example
Context	Reference something tangible — a launch, blog post, job detail, or user pattern	"I saw your team recently launched an orchestration layer for agent planning…"
Curiosity	Ask a "how," "why," or "what if" to invite dialogue	"…how do you decide which tools agents should have access to by default?"
Relevance	Connect it to your role or domain lens	"…especially in a PM role focused on trust and agent UX."

Figure 12.3: Structure for Crafting High-Signal Interview Questions

Apply this prompt formula

"I noticed [specific signal]. That made me curious about [insightful, role-relevant question]."

Examples:

- "I noticed your onboarding agents now include real-time document parsing. How do you decide what actions should be fully autonomous versus user-reviewed?"
- "Your recent post on agent orchestration mentioned tool chaining. What failure modes do you see most often—and how do you design around them?"
- "I read that your team's exploring retrieval-augmented generation (RAG) for summarization—what's been most challenging when aligning that with UX constraints in legal workflows?"

Each question does three things:

1. Shows you've done your homework

2. Anchors your thinking in real product decisions
3. Signals your fluency in system-aware PM thinking

Pro tips for maximizing signal:

- Don't ask what's easily Googleable—ask what only *they* can answer.
- Don't ask broad hypotheticals—anchor your question in a decision, tradeoff, or behavior.
- Don't monologue—use a crisp 10–15 second lead-in, then ask your question clearly.
- Match the question to the interviewer. Tailor your questions to the person across the table. With PMs or hiring managers, ask about roadmap priorities, metrics, and product tradeoffs. For engineers or researchers, focus on model behavior, orchestration, and system design. With executives or founders, explore strategy, vision, and competitive positioning. And with UX or design leads, ask about user trust, explainability, and intervention design. Each angle signals that you're thinking like a cross-functional partner.

Your questions are more than a formality—they're your final product moment. They show whether you think like someone who can drive tradeoffs, lead amid ambiguity, and own complexity.

Whether you're asking about model routing, intervention points, or success metrics for AI copilots, your questions should say:

> *"This candidate is already thinking like one of us."*

That's the strongest signal you can send.

Common pitfalls—And what to do instead

You've seen how to craft high-signal questions. Now here's the flip side—patterns that quietly weaken your signal. These aren't dramatic missteps, but subtle slips that can flatten an otherwise strong impression. Catch them early, and your final minutes won't just be polished, they'll be memorable.

Use Figure 12.4 below to spot—and sidestep—the most common traps.

Mistake	Why It Hurts	Better Move
Asking a question just to fill time	Feels shallow or unprepared — the interviewer senses it's not genuine	Prepare 2–3 specific questions tied to product decisions, org strategy, or your potential impact
Turning your question into a mini-monologue	Wastes time and shifts focus to you — not the dialogue	Keep context under 15 seconds, then pivot to a crisp, open-ended question
Asking GenAI questions that ignore the company's actual stack	Signals you don't understand their work — e.g., asking about agents when they're building infra	Research the team's product layer (infra, model, orchestration, app) and tailor accordingly
Avoiding risk, friction, or ethical questions entirely	Misses a chance to show judgment, systems thinking, and maturity	Ask respectfully: "What's been your approach to designing for failure modes or trust breakdowns?"
Focusing only on what the company does — not what you'd do there	You sound like a spectator, not a potential contributor	Ask about onboarding expectations, success metrics, or how your strengths could map to key goals

Figure 12.4: Common Mistakes and Best Practices for Asking Questions in GenAI PM Interviews

Final thought: Ask like a collaborator, Not a candidate

In GenAI product roles—especially those involving Agentic AI—your interview doesn't end with your last answer. It ends with how you ask, listen, and co-think.

Don't ask to impress. Ask to understand.
Don't signal curiosity. Model it.
Don't perform a role. Embody it.

Because at the end of the day, strong product managers aren't hired just for what they know—but for how they show up when things are undefined. And no part of the interview gives you more freedom to show that than this one.

The PREPARE framework gives you a complete, structured approach to confidently navigate every stage of your GenAI PM interview. From researching the company to asking high-signal questions, you now have the tools to lead with clarity, curiosity, and credibility. For additional topics like post-interview follow-ups, cultural fit, and offer negotiations, see Appendix E.

Chapter 13: Building the Future: A New Era for AI Product Managers

As we stand together at the close of this book, I want you to take a deep breath and look back at the path you've traveled.

When you first opened these pages, the world of Generative AI product management might have felt like a vast, untamed frontier—new technologies shifting beneath your feet, roles evolving faster than you could grasp, and big questions about ethics, impact, and opportunity looming large. You may have wondered: *Where do I even start? How do I build products if I'm not an engineer? What if I fall behind while the technology races ahead?*

These were not trivial questions. They are the very questions even seasoned leaders ask today. The fact that you chose to face them, to step forward into the unknown rather than shy away, already sets you apart.

Because now, you're not the same person who started this journey. Through these chapters, you haven't just gathered information. You've undergone a real transformation. You've built a foundation in the mechanics of GenAI—understanding how models work, how prompts can be tuned, and how success in AI systems must be measured with new forms of precision. You've wrestled with the ethical weight of building systems that can influence lives, economies, and societies, learning that responsibility isn't a constraint—it's a catalyst for better products. You've explored how to navigate the path into AI PM roles, how to show up in high-stakes interviews, and how to lead teams through ambiguity and change. You've begun to orchestrate the intricate dance between human creativity and AI intelligence, discovering that real innovation happens not in isolation, but through collaboration across models, tools, and people.

Most importantly, you've forged something deeper than knowledge. You've forged a mindset: one that thrives not on certainty, but on curiosity. A mindset that isn't shaken by change, but sharpened by it. A mindset that understands leadership in the AI era is not about commanding perfect answers, but about guiding bold questions, ethical experimentation, and rapid learning.

Because here's the deeper shift: GenAI isn't just another technology layer to bolt onto old ways of building. It is a fundamentally new way of creating products—where AI becomes not just a feature, but a collaborator; where success comes not from rigid control, but from dynamic orchestration.

In the past, product managers shaped tools. Today, you are shaping intelligent collaborators—systems that can perceive, reason, adapt, and interact with the world in ways no static application ever could.

And that demands a new kind of product maker.

In this new era, the most impactful AI product managers are not just strategic leaders. They are full-stack builders—willing to prototype ideas, chain together models and APIs, experiment hands-on with intelligent workflows, and learn fast by doing, not just by planning. They don't wait for innovation to happen. They spark it. They don't just ask, *"What can my team deliver?"* They ask, *"What can we create together?"*

You are stepping into that role now. Not just as a manager of projects, but as an orchestrator of systems. Not just as a student of AI, but as a builder, a shaper, a responsible leader of what's next.

The AI landscape will keep evolving. Models will improve. New techniques will emerge. Entire categories of products will rise and fall. Certainty will remain rare—but opportunity will remain endless. And your core strengths—curiosity, strategic thinking, full-stack experimentation, ethical leadership—will remain timeless.

You have journeyed from feeling overwhelmed by the speed of AI innovation to being equipped to build thoughtfully, responsibly, and creatively within it. You are no longer just learning to survive in a world shaped by AI. You are learning to shape that world yourself.

The future of AI product management is not yet written. But it will be written by builders like you—those who stay humble, stay curious, and keep moving boldly into the unknown.

And now, you carry the tools—and the mindset—to help write it.

Learning Capital Is the New Currency

In the world of AI, the most valuable thing you can accumulate isn't knowledge. It's *learning capital*—your ability to adapt, absorb, and grow faster than the world around you evolves.

The pace of change in this space defies traditional learning curves. New models emerge overnight. Capabilities shift quarterly. Tools rise, mature, and get replaced before most teams have even caught up.

There is no final playbook. No steady-state expertise.

That's why the most successful AI product managers won't be those who memorize frameworks or keep up with every release note. They'll be the ones who invest continuously in their learning capital—who treat curiosity like compounding interest, and who treat change not as a disruption, but as an invitation.

Your learning capital is what allows you to move with confidence even when the ground is shifting. It's what enables you to ask better questions in uncertain moments, to pivot quickly when a tool breaks, to take on new problems that have no precedent—and still lead. It's not just what you know—it's how quickly you can learn what's next.

The value of this capital is especially high in AI product management. Why? Because your job isn't just to manage timelines or ship features anymore. It's to guide dynamic systems—living products that evolve as they interact with the world, with users, and with other agents.

Products that don't behave deterministically. Products where your first launch is just the beginning of the learning loop.

Every model you touch, every prompt you tune, every orchestration flow you test becomes a data point—not just for the product, but for *you*. And your ability to turn that feedback into insight, into action, into improvement—that's your real compounding edge.

Learning capital also gives you leverage in the most human part of your job: your team. When your team sees you testing a new tool on your own, writing basic agent workflows to understand what's possible, or asking insightful questions about a model's edge case behavior, it inspires a culture where learning is normalized—not feared. That culture is your multiplier.

But this isn't just about staying current. It's about staying *valuable*. The future of AI product leadership belongs to those who never stop upgrading their perspective. Those who will flourish are those who continue to evolve their intuition. Who remain open, humble, and hungry, no matter how much they've already accomplished.

And unlike titles or technical expertise, learning capital isn't zero-sum. You don't have to wait for permission. You build it every time you pause to understand how something works, question why it behaves a certain way, or try building it yourself—even if it fails the first time. The AI world rewards motion, not mastery. It rewards iteration over certainty. It rewards the builder who says, *I don't know yet— but I'll figure it out.*

So keep moving. Keep learning.

And know that your greatest asset isn't the knowledge you hold—It's your capacity to learn faster, deeper, and more ethically than the world expects. Because in this economy of accelerated change, *learning capital is the new currency*—and you're already richer than you were when this journey began.

Your Superpowers as an AI PM

As you move forward in this ever-evolving landscape, you won't just be relying on tactics or tools. You'll be operating from a deeper foundation: three core superpowers that will fuel your journey as a full-stack product maker and lifelong learner. These aren't buzzwords. They're the essential capabilities that will define how you show up—on your teams, in your products, and in your impact on the world.

The first is **a**daptation.

The AI landscape is a living, breathing organism. Models shift, needs change, frameworks evolve, and what works today may become irrelevant tomorrow. Your superpower lies not in resisting change—but in learning faster than the world transforms. You'll need to stay open to new ideas, new architectures, and new ways of thinking. When a prompt behaves unpredictably or a team hits a roadblock, you'll ask: *What's this trying to teach me?* And in doing so, you'll create a culture where growth is a shared adventure, not a personal burden.

The second is **o**rchestration.

In traditional PM work, your job might have been to define and ship features. But in the GenAI world, your job is to orchestrate—models, humans, systems, data, and trust—all into intelligent, adaptive, reliable experiences. You're a conductor, not just a planner. You'll be the one pulling together messy tools and fragile connections and turning them into elegant user flows. You'll experiment with agents, connect APIs, trace outputs, debug failures—and through that hands-on work, you'll master the new art of intelligent systems design.

And finally, there's e*thical leadership.*

Because in this space, your work doesn't just scale functionality—it scales influence. The products you help build may one day guide decisions, shape beliefs, or even redefine social norms. That means your responsibility doesn't stop at the feature spec. You must ask: *Can we build this?*—but more importantly: *Should we?* You'll lead by staying curious about the downstream impact of every choice— seeking insight not just from users, but from communities, edge cases, regulators, and your own values. Ethics, in this era, isn't a compliance box. It's a differentiator. It's a trust multiplier. It's the soul of your product.

These superpowers—adaptation, orchestration, and ethical leadership—are not optional.

They're what enable you to navigate chaos without losing clarity. They're what help you build at the frontier without compromising what matters. They are your compass as you help define what product leadership means in the age of AI. And they're not static traits. They grow with you. The more you learn, the stronger they become. The more you practice them, the more naturally they guide your work. So as you move forward, don't just ask, *What should I do next?* Ask, *Which of my superpowers do I need to strengthen right now?*

Because these are what will carry you through the noise, the novelty, the breakthroughs, and the failures. They will help you build products that don't just function—they inspire. They connect. They endure.

The Future You'll Help Shape: Imagine the world you're stepping into.

Picture a healthcare app that co-creates treatment plans with patients, adapting in real time to their symptoms, preferences, and emotional needs. Envision a learning platform powered by AI agents that personalize education for millions, bridging gaps across languages and cultures. Dream of decentralized AI ecosystems that enable small businesses in Nairobi, farmers in São Paulo, and innovators in Seattle to collaborate seamlessly, sharing insights and resources in ways we can barely fathom today.

These aren't distant fantasies—they're the seeds of a future you'll help grow. You won't just build tools; you'll craft intelligent collaborators that redefine how we live, work, and dream. Imagine AI systems that monitor their own fairness, flagging biases before they take root. Picture products that anticipate user needs—not just responding, but inspiring users to think bigger, create more, and connect deeper. This is the world you're shaping, and your commitment to learning will be the thread that ties it all together.

As an AI PM, you'll face moments of uncertainty—when a model fails, when a user's trust is at stake, when the technology outpaces the policies. But those moments are also opportunities to learn, to innovate, to lead with heart. Your curiosity will help you turn challenges into breakthroughs, building products that don't just solve problems—they spark possibility.

Final Words: Keep Learning, Keep Building

This book was never about handing you a finished puzzle. It was about giving you the pieces—strategies, frameworks, mindsets—and trusting you to build something extraordinary. You've got the tools to navigate the AI era, the vision to see what's possible, and the heart to make it meaningful. Now, it's time to keep going.

The story of AI product management is still being written, and you're one of its authors. Stay humble, even when the path feels uncertain—every question is a chance to grow. Stay bold, even when the challenges feel big—your ideas can change the world. Stay curious, always learning from the users, teams, and technologies around you.

Think of the PMs who inspire you—the ones who built products that touched lives, who dared to dream beyond the status quo. You're part of that lineage now.

So, go out there and build wisely. Build bravely. Build with a learner's heart. Ask yourself: *What can I learn today? What can I create tomorrow?* The answers will lead you to places you can't yet imagine. The world is waiting for the possibilities you'll bring to life.

Keep learning, keep building, and never stop shaping what's next.

Appendix

Appendix A: Glossary

This glossary introduces foundational terms that are important for understanding how Gen AI and Agentic AI systems function. These concepts will help product managers communicate effectively with technical teams and make informed product decisions when working with AI-powered systems.

Agentic AI: An advanced form of AI that exhibits goal-directed behavior, autonomy, adaptability, and the ability to coordinate multiple agents or tools to complete complex tasks with minimal human intervention.

AI Agent: A software entity that performs specific tasks autonomously in response to user input or environmental triggers. Typically limited in scope and designed for task execution (e.g., chatbots, scheduling assistants).

API (Application Programming Interface): A set of ready-made "doors" that one program uses to talk to another—send a request in, get a response back. In GenAI and Agentic AI, APIs are how your product calls a model (e.g., "generate a summary") or plugs in extra tools without rebuilding everything from scratch.

Context Window: The maximum number of tokens a model can consider in one prompt-response cycle; exceed it and you must truncate, chunk, or summarize, which shapes choices in RAG and agent memory.

Copilot: An AI assistant embedded within a software application that helps users complete tasks using natural language (e.g., Microsoft 365 Copilot).

CUDA system (Compute Unified Device Architecture): NVIDIA's software-and-hardware stack that turns a graphics card (GPU) into an AI workhorse. CUDA lets thousands of tiny calculations run in parallel, speeding up the math behind training and running GenAI models and agent workflows.

Data Drift: Changes in an AI system's behavior over time due to shifts in the data it processes, which can affect its accuracy or fairness and needs to be monitored.

Explainability: Tools or techniques that help users understand why an AI model made a particular decision or prediction.

Fine-Tuning: The process of adapting a pre-trained foundation model to a specific domain or task using additional data.

Foundation Model: A large, pre-trained AI model that can be adapted to many tasks, such as text generation, image analysis, or coding.

Four-Dimension Evaluation Framework: A straightforward way to judge a GenAI system from four angles—User (adoption, task-completion, satisfaction, trust), Model (accuracy, groundedness, consistency), System (latency, uptime, tool-call success, cost efficiency), and Business (revenue, retention, time-to-value, compliance)—so product managers see a complete picture instead of over-optimizing one area at the expense of the others.

Four-Lens Framework: A framework for evaluating GenAI opportunities to confirm they are both meaningful and viable—beginning with the deep user problem (frustration), checking fit with GenAI's unique strengths, weighing feasibility and risk, and ensuring behavioral and UX alignment.

Generative AI Model: A type of AI model that produces new content—such as text, code, images, or audio—based on patterns learned from training data. Unlike traditional AI systems that classify or retrieve data, generative models are capable of generating original outputs. Examples include GPT-4, Claude, and DALL·E.

GPU (Graphics Processing Unit): A specialized processor designed to accelerate complex computations, widely used in training and running AI models.

Graph RAG: An advanced RAG pattern using graph-based knowledge for relational reasoning, often in research or analytics.

Grounding (Groundedness): The practice of anchoring model outputs in verified source data—such as company docs or real-time search—so responses stay factual and reduce hallucination.

Guardrails: Rules and limits in a GenAI system to prevent misuse, harmful outputs, or overreach, such as rate limits, authentication, or prompt filters.

Hallucination: When the model confidently generates factually incorrect or misleading responses. Common in generative systems.

Human-in-the-Loop (HITL): A safeguard in which people review, correct, or approve AI outputs—essential when stakes are high or data is sensitive.

Hyperparameter Tuning: Optimizing a model's learning settings (e.g., learning rate, batch size) to improve performance and avoid overfitting. Use it after model selection to enhance accuracy (e.g., tuning a model for better demand prediction).

Inference: The process of generating output from the trained model using new input. Happens in real time as the model predicts the next token in a sequence.

Inference Endpoint: A hosted access point that allows users to send input to a model and receive output, often via an API.

LoRA (Low-Rank Adaptation): A resource-efficient fine-tuning method for large models, modifying only low-rank matrices to reduce computational cost. Use it for large models when full fine-tuning is too expensive (e.g., fine-tuning Llama with limited resources).

MCP: A protocol for advanced GenAI systems enabling universal knowledge access and cross-system integration

Memory (in AI Agents): A framework that enables AI agents to store and recall information over time. This may include previous steps taken in a task, prior user inputs, or other contextual data. Memory allows agents to operate more coherently across multiple interactions or long workflows.

Model Deployment: The act of putting an AI model into a live environment where users or applications can interact with it.

Model Drifting: When production data deviates from training data, causing performance decline. Monitor for drift in dynamic environments (e.g., financial markets) and retrain as needed (e.g., updating a stock prediction model).

Orchestration Framework: A toolkit (like LangChain or Semantic Kernel) that coordinates how AI agents plan tasks, access tools, and retrieve knowledge.

Planning Framework: A system that allows an AI agent to break down a high-level goal into smaller, actionable steps. Planning frameworks enable the agent to decide the sequence of actions needed to complete a task and dynamically adjust the plan based on new inputs or outcomes.

Playgrounds: Interactive environments for experimenting with pre-trained models, testing prompts, and fine-tuning parameters. Use them for rapid prototyping and customization (e.g., testing prompts in a GPT playground).

Prompt Tuning: A lightweight way to customize a pre-trained GenAI model by learning a tiny set of special "prompt tokens" that steer the model's replies toward a specific task or style. It leaves the model's core weights untouched, making it faster and cheaper than full fine-tuning and easy to swap out for different use cases.

Prompt Engineering: The iterative craft of designing and refining prompts to steer a GenAI model toward the desired output without changing its weights.

Prompt Filters: Guardrails that block harmful or inappropriate inputs to a GenAI system, such as hate speech or medical advice, to ensure safe operation.

Prompt Leakage: When a GenAI system accidentally exposes information from prior prompts or past sessions in new responses—especially sensitive or user-specific content.

Prompt Versioning: The practice of maintaining different versions of prompts to enable A/B testing, rollback, or updates without disrupting system behavior.

Reinforcement Learning: A machine learning approach in which an agent learns by interacting with its environment and receiving feedback in the form of rewards or penalties. Reinforcement learning is often used to train AI agents to make decisions in dynamic or complex settings, such as robotics or autonomous navigation.

Responsible AI: A set of practices that ensure AI systems are safe, ethical, fair, and compliant with laws and societal norms.

Retrieval-Augmented Generation (RAG): A technique where AI models fetch external knowledge (e.g., documents, databases) to enhance their output accuracy.

SDK (Software Development Kit): A ready-to-use collection of libraries, sample code, and documentation that lets developers plug a platform's features into their app with minimal setup. In GenAI and Agentic AI, an SDK typically handles tasks like sending prompts to a model, streaming back responses, and managing authentication—so teams can focus on product ideas instead of low-level wiring.

Tokens: Smallest units of input text processed by an AI model. They can be parts of words, whole words, or punctuation. Used to compute model cost and manage input length.

Tokenization: The process of breaking text into tokens before processing. Essential for model comprehension and cost calculation.

Tool Use (APIs, Plug-ins, Retrieval): Agentic AI extends its reach by calling external APIs, invoking software plug-ins, or fetching data through retrieval mechanisms, giving it real-time information and the ability to act beyond what the model alone can do.

TPU (Tensor Processing Unit): A custom AI accelerator developed by Google to speed up machine learning tasks, especially for large-scale models.

Training Data: The massive corpus (text, code, images) used to train AI models. It teaches the model language patterns, factual knowledge, and associations.

Transformer: A neural network architecture introduced in 2017 that enables deep contextual understanding. Most modern LLMs use this architecture.

Self-Attention: A mechanism inside Transformers that lets the model weigh the importance of each token in a sentence relative to others. Critical for understanding language nuance.

System Prompt: A behind-the-scenes instruction that establishes the model's persona, rules, or brand voice before user prompts arrive, helping enforce consistency and safety.

Parameters: The "learned memory" of a model—stored as numeric values across the neural network. More parameters usually mean higher capacity for reasoning and generation.

Embeddings: Numerical representations of words, phrases, or concepts in multi-dimensional space. They allow AI to understand similarity and meaning.

Vectors: Mathematical objects derived from embeddings that allow similarity-based search and retrieval. Used in RAG systems with vector databases.

Vector Database: A system like Pinecone or Weaviate that stores and retrieves embeddings quickly to provide relevant data to the model in real time.

RAG (Retrieval-Augmented Generation): A technique where external documents are retrieved (using vector similarity) and passed to the model for more accurate responses.

Appendix B: Design considerations for interaction patterns

Small design details will make the difference between an AI that feels supportive and one that feels confusing or overbearing. This tuning is where Principle 2 (Intuitive Interaction), Principle 4 (Trust and Safeguards), Principle 6 (Learning and Feedback), and Principle 7 (Inclusive Access) come to life. Here are the key considerations to guide you:

Clarity—users should immediately understand how to engage with the AI:

- If you choose a chat-based pattern for a high-oversight task, make sure prompts are simple, predictable, and discoverable.
- If you choose a copilot pattern, clearly signal when the AI is available—a subtle icon, a soft highlight, or a status indicator—so users know help is there without feeling pressured.

Clear entry points reduce friction and build confidence. Control balance—match the AI's autonomy to user expectations:

- In copilot models, users must feel firmly in charge — able to ignore or edit suggestions easily.
- In agentic systems, always provide visible controls like "Pause," "Cancel," or "Undo"—even if the task is low-oversight.

Autonomy without visible user control is one of the fastest ways to erode trust. Accessibility design for diverse users from the start:

- Offer flexible input modes (voice, text, visual cues).
- Ensure accessibility compliance across languages, abilities, and tech-savviness.

Building inclusivity isn't just good ethics—it's good product sense. This ties directly to Asha Sharma's philosophy: the fastest-growing AI products are those accessible to the broadest user base. Feedback loops—create lightweight ways for users to give feedback:

- A simple thumbs-up/down.
- A prompt like "Was this helpful?"
- A fast way to report hallucinations or misbehavior.

Continuous learning isn't just for AI. It's for the PM team too.

Appendix C: Detailed Metrics for GenAI Evaluation

Evaluating GenAI products requires a holistic view — one that balances model performance, user experience, business goals, and system reliability.

This appendix extends the core metrics introduced in Chapter 6 by providing a comprehensive set of evaluation criteria organized by the Four-Dimension Evaluation Framework: 1) User Dimension – Experience and engagement 2) Model Dimension – Output quality and safety 3) System Dimension – Infrastructure and scalability 4) Business Dimension – Value, impact and risk. Each lens includes subcategories and practical metrics that you can adapt based on your product's goals and maturity.

User Dimension Metrics

Focuses on how end-users interact with the GenAI system. Is the experience useful, trustworthy, and satisfying?

Engagement metrics
- Adoption Rate: Percentage of eligible users actively using the feature
- Frequency of Use: How often users return to the GenAI experience
- Session Length / Queries per Session: Measures depth of interaction
- Query Length: Indicator of user effort or intent granularity
- Thumbs Up/Down Feedback: Quick sentiment gauge on usefulness
- Satisfaction Scores: From surveys, Likert scales, or in-app feedback

Efficiency metrics
- Task Completion Rate: Did the user succeed using the AI feature?
- Time to Completion: How long it took to get value
- Fallback Rate: Percentage of users who abandon the AI and use alternatives (e.g., manual search)

Model Dimension Metrics

These metrics evaluate the AI model's performance, focusing on output quality, safety, and comparative evaluation.

Output Quality
- Accuracy/Precision/Recall /F1 Score: Useful for structured tasks (e.g., retrieval, classification)
- Coherence: Logical flow of the model's output
- Fluency: Naturalness of language or interface interactions
- Groundedness: Output aligns with the retrieved or source data
- Instruction Following: Model compliance with user instructions
- Conciseness vs. Verbosity: Balance between detail and clarity
- Summarization Quality: For tasks involving long-form content

Safety and Compliance
- Toxicity Score: Frequency of harmful or offensive outputs
- Guardrail Trigger Rate: How often safety filters are activated
- Prompt Leakage Rate: Whether prior prompts bleed into future responses
- Bias Metrics: Bias detection across sensitive dimensions (e.g., gender, race)
- Error Type Distribution: Classifying hallucinations, refusals, or irrelevant answers

System Dimension Metrics

These metrics examine the technical infrastructure, focusing on deployment, reliability, throughput, and cost.

Deployment & Monitoring
- Model Time to Deployment: Speed from development to production
- Number of Deployed Models: Indicates scale or portfolio breadth
- Percent of Models with Active Monitoring: Observability coverage
- Automated Pipeline Coverage: Share of deployments that are fully automated

Reliability & Responsiveness
- Latency (p95, p99): How quickly the model responds
- Uptime %: System availability over time
- Error Rate: Percentage of failed or incomplete responses

Throughput & Utilization

- Request Throughput (req/sec): Number of real-time requests handled
- Token Throughput (tokens/sec): Volume of tokens processed
- Serving Node Utilization: Infrastructure usage efficiency
- GPU/TPU Utilization: e.g., B200 NVL72 or MI300X clusters

Cost

- Cost per 1 M Tokens – Normalised usage pricing
- Serving Cost per Session / User – Pricing insight
- Cost-to-Value Ratio – Cost vs. key business metric
- Energy per 1 M Tokens (J) – Efficiency benchmark
- CO_2-eq per Interaction – Carbon footprint signal

Business Dimension Metrics

These metrics connect the AI to organizational goals, focusing on financial impact, customer growth, and operational resilience.

Financial Impact

- Cost Savings: Reduction in operational effort or headcount
- Time Saved per User: Especially in productivity-enhancing products
- Automation ROI: Value created by shifting manual tasks to GenAI

Customer Growth

- New User Conversion: Is GenAI attracting new users or markets?
- Customer Retention Lift: Do AI features improve stickiness?
- Upsell Impact: Correlation between AI usage and tier upgrades

Innovation & Expansion

- New Use Case Enablement: Did GenAI unlock capabilities not possible before?
- Net New Revenue: From AI-enabled products or features
- Partner Ecosystem Value: For platforms or marketplaces

Trust and Risk Mitigation

- User Trust Index: Composite metric from surveys, usage patterns, and sentiment
- Regulatory Readiness – Alignment with frameworks like the EU AI Act (high-risk provisions enter phased enforcement Q2 2026)
- Security Incident Rate: Model or system vulnerabilities

Note: You don't need to track every metric—pick the ones that matter for your product's phase, risk level, and target outcomes. As a GenAI PM, your job isn't just to measure success. It's to define it clearly—and ensure your system evolves to meet that bar.

Appendix D: Transition Paths to GenAI PM by Background

Not everyone begins their journey to GenAI Product Management from the same starting point—and that's exactly the point. This appendix is your tactical guide to navigating into GenAI PM based on where you are today.

Whether you're a software engineer, a UX designer, a marketer, or a student, the key is to identify your zone of strength and build intentionally from there. Think of these not as rigid tracks, but as launchpads. Each starting point leads to new expansion zones, and with the right strategy, every role has a path in.

Developers/Data Scientists

Zones of strength: Infra Mountains, AI Foundations, Evaluation Range
Example entry roles:

- Platform product manager (AI services, model hosting)
- Model evaluation or fine-tuning PM
- GenAI Infrastructure PM

What to leverage:

- Deep technical fluency with AI/ML tools and APIs
- Familiarity with training pipelines, MLOps, and cloud infrastructure
- Hands-on exposure to frameworks like Hugging Face, LangChain, Semantic Kernel, or OpenAI APIs

What to build:

- Product instincts: translating user pain points into system design
- Cross-functional leadership and stakeholder communication
- Strategic thinking across the AI value chain

Your edge: You already speak the technical language—now use it to bridge engineering with user and business value.

UX Designers / UX Researchers

Zones of strength: user coast, prompt plains
Example entry roles:

- Human-AI interaction PM
- Prompt engineering PM
- Responsible AI experience PM

What to leverage:

- User empathy and trust-building instincts
- Experience shaping complex user flows and interaction design
- A track record of simplifying complex systems for users

What to build:

- Working knowledge of GenAI systems (e.g., hallucination, model drift, prompt iteration)
- Comfort collaborating with infra and research teams
- Ability to translate model behavior into intuitive, trustworthy interfaces

Your edge: You think in experience, not just features. That's exactly what GenAI teams need to craft trusted interactions.

Product Marketing Managers/Business Strategists

Zones of strength: Biz Model Island, PM Farmlands
Example starting roles:

- Monetization strategy PM (GenAI pricing & packaging)
- Go-to-market (GTM) PM for GenAI models
- Vertical solutions PM (AI in healthcare, finance, retail, etc.)

What to leverage:

- Market positioning and pricing acumen
- Strength in crafting compelling narratives and product messaging
- GTM execution and adoption strategies

What to build:

- Technical fluency around GenAI APIs and constraints
- Comfort navigating model capabilities, inference costs, and feedback loops
- Ability to align GTM strategy with technical feasibility

Your edge: You can bring AI products to market with clarity and commercial rigor. GenAI teams need your commercial instincts.

Traditional PMs Without AI Background

Zones of Strength: PM Farmlands → any adjacent zone
Example starting roles:

- PM for AI-enhanced features
- Copilot extensions PM
- User workflow automation PM

What to leverage:

- Core product skills: roadmap ownership, delivery discipline, customer insight
- Strength in navigating ambiguity and cross-functional alignment

What to build:

- GenAI system literacy: prompts, model lifecycle, failure modes
- Understanding of where GenAI changes traditional product dynamics
- Confidence working across engineering, research, and design

Your edge: You already know how to ship. Now you just need to learn how GenAI changes the rules of the game.

AI Practitioners (ML Engineers, Research Scientists)

Zones of strength: AI Foundations, Infra Mountains, Evaluation Range
Example starting roles:

- Model-to-product PM
- AI evaluation & observability PM
- Research transition PM

What to leverage:

- Deep familiarity with model development, tuning, and evaluation
- Technical authority and firsthand awareness of trade-offs

What to build:

- Customer framing: what users need, not just what models can do
- Communication skills to align researchers, PMs, and business stakeholder

- Product strategy and ownership mindsets

Your Edge: You already shape the models—now shape the products that use them.

Students/Career Starters

Zones of strength: Data Plains, PM Farmlands (early seeds)
Example starting roles:

- Associate product manager (GenAI or AI-adjacent teams)
- AI product analyst
- Program manager for AI pilots

What to leverage:

- Curiosity and flexibility to explore different parts of the stack
- Exposure to modern tools and GenAI experimentation
- Energy and openness to iterate quickly

What to build:

- Small but real projects (GitHub, case studies, prompt demos)
- Public proof of your learning journey (LinkedIn, blogs, portfolios)
- Mentorship and internship opportunities that offer access to GenAI product teams

Your edge: You're early—which means you can learn broadly and move fast. That's a superpower in a fast-moving field.

Domain Experts (Healthcare, Finance, Legal, etc.)

Zones of strength: Ethical Forest, Biz Model Island, PM Farmlands
Example starting roles:

- Industry AI product manager
- Responsible AI PM (regulated use cases)
- Domain copilot PM (e.g., LegalCopilot, HealthCopilot)

What to leverage:

- Subject-matter depth and regulatory context
- Intuition for risk, trust, and user expectations in your field
- Awareness of real-world pain points and unmet needs

What to build:

- System-level GenAI literacy and technical framing
- Cross-functional collaboration with AI teams
- Translational storytelling from domain → model behavior

Your edge: GenAI needs grounding in real-world constraints—and you bring the context to make it useful and trusted.

Appendix E: Beyond PREPARE—Additional Topics for Interview Success

Once you've completed the core PREPARE framework—from researching the company to rehearsing Q&A—there are still a few strategic areas that can sharpen your edge as a GenAI Product Manager candidate. Think of this appendix as your bonus playbook: these aren't always discussed in depth, but they can meaningfully influence outcomes, especially in a competitive GenAI hiring market.

1. Handling Take-Home Assignments

Many companies ask candidates to complete a short product case or design doc to evaluate structured thinking, AI awareness, and communication clarity.

- Clarify scope early: Are you expected to cover UI/UX, model choices, metrics, or just product flows?
- Define assumptions: Call out data constraints, user behaviors, or model limitations.
- Keep it tight: A concise, well-structured 1–2 pager or 10-slide deck is often more effective than a lengthy document.
- Balance business and system thinking: Show how user needs, technical constraints, and ethical risks intersect — and how you'd navigate them.

2. Demonstrating Team Fit & Cross-Functional Synergy

Your success in GenAI PM roles depends on trust and collaboration across highly specialized teams — modelers, researchers, infra, compliance, and UX.

- Reflect on how you've worked with adjacent teams under ambiguity.
- In your answers or follow-up questions, ask how PMs partner with researchers, data scientists, or UX for evaluation, alignment, or drift recovery.

- Show that you're not just a builder — you're a bridge.

3. Offer Negotiation & Decision-Making

GenAI PM roles often include equity, unique benefits (e.g., GPU access, conference stipends), and cross-functional ownership. Don't default to base salary comparisons.

- Do your research: Use Levels.fyi, teamblind.com, or conversations with peers.
- Clarify value beyond comp: Ask about scope, impact, support for up-skilling, or involvement in foundational bets (e.g., orchestration or agentic product layers).
- Negotiate thoughtfully: Frame it as a win-win — what you'll bring and what you need to succeed.

4. On-Site and Final Round Nuances

Whether virtual or in person, final loops are your chance to demonstrate product judgment, adaptability, and culture fit — not just knowledge.

- Whiteboard sessions: Be ready to sketch data flows, agent loops, or feedback circuits — clearly and with tradeoffs.
- Lunch or casual chats: These may be informal but are still evaluative. Stay curious, personable, and grounded in your product mindset.
- Observe team dynamics: Notice communication styles, role clarity, and how they talk about failures or feedback.

5. Personal Brand & Thought Leadership

For senior PM roles, especially in early-stage or high-visibility GenAI teams, your public work — blog posts, talks, or open-source contributions — can signal credibility and influence.

- Curate your online presence: Highlight relevant case studies, teardown threads, or learnings from experiments.
- Reference your work naturally in interviews if applicable.
- Stay authentic: You don't need to be loud — just thoughtful, helpful, and signal-aligned.

6. Inside the Recruiter "Algorithm"

Guest Insight by Jeremy Schifeling—former recruiter and hiring manager at LinkedIn and Khan Academy, and author of Career Coach GPT, the #1 AI for careers best-seller.

So you're a master of AI algorithms…but do you understand the most important algorithm of all: the human one?

As much as hiring processes leverage AI today, the final decisions still come down to the judgment of two humans: the recruiter and the hiring manager. Here's a crash course on those human black boxes:

1. Human hiring "algorithms" are risk-averse. No recruiter wants to get chewed out for championing a candidate who makes no sense, just as no hiring manager wants to tie their career to someone unproven.

2. Your #1 goal is to de-risk your candidacy—make yourself the safe, slam-dunk choice. Think "No CTO ever got fired for buying IBM." Likewise, no recruiter gets in trouble for selecting the obvious candidate.

3. The easiest way to make yourself an obvious candidate is to make yourself an insider. That means that:
   a. If you're breaking into AI PM for the first time, use the industry's language on your résumé, cover letter, and LinkedIn profile—leveraging Saumil's AI keyword hacks from Chapter 11.

   b. If you're early in your career, show a proven track record of success by quantifying impact in classes, extracurriculars, or volunteer work. When I worked at LinkedIn, we surveyed 1,000 recruiters and discovered that 40 % view volunteer work on par with paid experience!

   c. If you're brand-new to the industry, build connections on the inside. Use LinkedIn to find alumni from your alma mater, employees from your former company, or 2nd-degree connections who can introduce you to the hiring manager—and even go to bat for you.

Bottom line: definitely keep mastering powerful AI algorithms. But don't forget about the human ones—at least until AI takes over hiring too! Final Reminder.

Appendix F: How to Stay Updated

The world of generative AI moves fast—and so does the role of the AI product manager. To help you keep pace with new frameworks, tools, and hiring practices, I publish a Substack newsletter, delivered to your inbox on the first Monday of every month.

What You'll Get:

- Fresh content updates as frameworks, tools, and trends evolve
- Curated interview prep tips and GenAI-specific PM interview questions
- Peer discussion threads with others navigating the same transitions
- Behind-the-scenes stories from PMs switching into GenAI roles
- Bonus resources I didn't include in the book (because they change too fast!)

How to Join:

1. Visit **zerotogenaiproductleader.substack.com**
2. Click **Subscribe** and choose the free tier
3. To unlock your complimentary access (for book owners):
    a. Forward your **purchase confirmation email** (showing your name, date, and order)
    b. Send it to **abirllcservices@gmail.com** with the subject line **"Newsletter Access"**

The book is your launchpad; the newsletter is your continuous-learning runway. I'll see you there.

References

1. Abrego, Michelle. 2024. "AI Is Core to JPMorgan's $18 Billion Tech Investment. Here's What Its Execs Revealed About How It's Reshaping the Bank." *Business Insider*, November 2024. https://www.businessinsider.com/jpmorgan-how-artificial-intelligence-transforming-workflows-efficiencies-2025-5.
2. Abridge. *"CHRISTUS Health Cuts After-Hours Documentation Time by Over 60% with Abridge's AI Medical Scribe."* Abridge.com. Accessed October 2024. https://www.abridge.com/press-release/christus-health-announcement
3. Allen & Overy. n.d. "AI at Work: Partnering with Harvey." Accessed April 22, 2025. https://www.allenovery.com.
4. Anthropic. "Claude Pro: More Power, Priority Access." September 7, 2023. https://www.anthropic.com/news/claude-pro.
5. Anthropic. "Claude in Slack: Enterprise-Ready, Safe AI Assistance." Accessed April 22, 2025. https://www.anthropic.com/claude-in-slack.
6. Banks, Jeanine, and Tris Warkentin. *"Introducing Gemma: Open Models Built for Responsible AI Development."* Google Blog. February 21, 2024. https://blog.google/technology/developers/gemma-open-models/.
7. Bass, Dina. "Anthropic Says Gaining Traction with Finance and Health-Care Firms." Bloomberg, March 14, 2024. https://www.bloomberg.com/news/newsletters/2024-03-14/anthropic-says-gaining-traction-with-finance-and-healthcare-firms.
8. BBC. 2024. "Glue Pizza and Eat Rocks: Google AI Search Errors Go Viral." *BBC News*, May 24, 2024. https://www.bbc.com/news/articles/cd11gzejgz4o.
9. Brookings Institution. 2023. "AI in the Global South: Opportunities and Challenges Towards More Inclusive Governance." *Brookings Institution*. https://www.brookings.edu/articles/ai-in-the-global-south-opportunities-and-challenges-towards-more-inclusive-governance/
10. BusinessWire. 2024. "Grammarly Announces Tools to Measure the ROI of AI." *BusinessWire*, October 21, 2024. https://www.businesswire.com/news/home/20241021109558/en/Grammarly-Announces-Tools-to-Measure-the-ROI-of-AI.
11. CAF. 2025. "Artificial Intelligence (AI) at the Service of Financial Inclusion." *CAF Development Bank of Latin America*. https://www.caf.com/en/blog/artificial-intelligence-ai-at-the-service-of-financial-inclusion/
12. Chen, Caiwei. "Everyone in AI Is Talking About Manus. We Put It to the Test." *MIT Technology Review*, March 11, 2025. https://www.technologyreview.com/2025/03/11/1113133/manus-ai-review/
13. Chen, May Yee. 2024. "Taiwan Hospital Deploys AI Copilots to Lighten Workloads for Doctors, Nurses and Pharmacists." *Microsoft Source Asia News*, July 12, 2024. https://news.microsoft.com/source/asia/features/taiwan-hospital-deploys-ai-copilots-to-lighten-workloads-for-doctors-nurses-and-pharmacists/.

14. Chui, Michael, Roger Roberts, Lareina Yee, Eric Hazan, Alex Singla, Kate Smaje, Alex Sukharevsky, and Rodney Zemmel. 2023. "The Economic Potential of Generative AI: The Next Productivity Frontier." *McKinsey & Company*, June 14, 2023. https://www.mckinsey.com/capabilities/mckinsey-digital/our-insights/the-economic-potential-of-generative-ai-the-next-productivity-frontier.
15. CNBC. 2025. "Nvidia Turned CoreWeave into a Major Player in AI Years Before Helping to Save Its IPO." March 30. https://www.cnbc.com/2025/03/30/coreweaves-7-year-journey-to-ipo-wound-through-crypto-before-ai.html.
16. CoreWeave. 2025. "CoreWeave Partners with IBM to Deliver a New AI Supercomputer for IBM Granite Models." January 15. https://www.coreweave.com/news/coreweave-partners-with-ibm-to-deliver-new-ai-supercomputer-for-ibm-granite-models. .
17. DeepSeek AI. *"DeepSeek LLM Series."* Accessed October 2024. https://www.deepseek.com/.
18. Drenik, Gary. "Small AI Firms Challenge Tech Giants, Spurring Innovation." *Forbes*, April 17, 2025. https://www.forbes.com/sites/garydrenik/2025/04/17/small-ai-firms-challenge-tech-giants-spurring-innovation/.
19. Duolingo. 2023. "How Duolingo Uses AI." *Duolingo Blog.* Accessed Jan 2024. https://blog.duolingo.com/how-duolingo-uses-ai/.
20. Eckel et al. 2025 ""GenAI in Ecosystem Dynamics: Optimizing Product Teams." *Boston Consulting Group*, April 2025. https://www.bcg.com/x/the-multiplier/genai-in-ecosystem-dynamics-optimizing-product-teams.
21. Economic Times. 2025. "TCS to Add AI Agents alongside Human Workforce: N. Chandrasekaran." *The Economic Times*, April 2025. https://economictimes.indiatimes.com/tech/artificial-intelligence/tcs-to-add-ai-agents-alongside-human-workforce-n-chandrasekaran.
22. Etori, Naome, Ebasa Temesgen, and Maria Gini. 2023. "What We Know So Far: Artificial Intelligence in African Healthcare." *arXiv.* https://arxiv.org/abs/2305.18302
23. European Commission. 2024. "The EU Artificial Intelligence Act: Timeline and Key Provisions." *Official Journal of the European Union*, July 12. Accessed January 2025. https://eur-lex.europa.eu/oj/direct-access.html.
24. Fidelity Investments. 2022. "Fidelity Investments® Launches Innovative New Regulatory Technology (RegTech) Business to Help Financial Institutions Create Compliant Public Communications." Press release, January 24, 2022. . https://newsroom.fidelity.com/pressreleases/fidelity-investments--launches-innovative-new-regulatory-technology--regtech--business-to-help-finan/s/b776e743-d2cc-4f46-a730-1ae9cb5d2e07
25. Figma. 2024. "Meet Figma AI: Empowering Designers with Intelligent Tools." *Figma Blog.* June 26. https://www.figma.com/blog/introducing-figma-ai/.
26. Freger, Jonathan. 2023. "How Consumer Distrust of GenAI Impacts Its Implementation." *Forbes Technology Council*, November 21, 2023.

https://www.forbes.com/councils/forbestechcouncil/2023/11/21/how-consumer-distrust-of-genai-impacts-its-implementation/.

27. Gartner. 2024. "Gartner Identifies the Top 10 Strategic Technology Trends for 2025." October 21. https://www.gartner.com/en/newsroom/press-releases/2024-10-21-gartner-identifies-the-top-10-strategic-technology-trends-for-2025.

28. GitHub. 2022. "Research: How GitHub Copilot Helps Improve Developer Productivity." *GitHub Blog*. July 14. https://github.blog/news-insights/research/research-how-github-copilot-helps-improve-developer-productivity/.

29. GitHub. 2024a. "Quantifying GitHub Copilot's Impact in the Enterprise (with Accenture)." The GitHub Blog, May 9, 2024. https://github.blog/news-insights/research/research-quantifying-github-copilots-impact-in-the-enterprise-with-accenture/

30. GitHub. 2024b. *GitHub Universe 2024*. Accessed June 2025. https://githubuniverse.com/.

31. GitHub. n.d. "GitHub Copilot: Your AI Pair Programmer." Accessed Feb 4, 2025. https://github.com/features/copilot.

32. GitHub. n.d. "Reviewing User Activity Data for Copilot in Your Organization." *GitHub Docs*. Accessed March 6 2025. https://docs.github.com/en/copilot/how-tos/administer/organizations/reviewing-activity-related-to-github-copilot-in-your-organization/reviewing-user-activity-data-for-copilot-in-your-organization.

33. Goldman Sachs. 2023. "Generative AI Could Raise Global GDP by 7 Percent." *Goldman Sachs Insights*, April 5. https://www.goldmansachs.com/insights/articles/generative-ai-could-raise-global-gdp-by-7-percent.html.

34. Goldman, Sharon. 2025. "A Customer Support AI Went Rogue—and It's a Warning for Every Company Considering Replacing Workers with Automation." *Fortune*, April 19, 2025. https://fortune.com/article/customer-support-ai-cursor-went-rogue/

35. Google. 2019. "Book a Table with the Google Assistant across the Country on More Devices." *The Keyword* (blog). March 6. https://blog.google/products/assistant/book-table-google-assistant-across-country-more-devices/.

36. Google. 2024. "Measuring GenAI Success: A Deep Dive into the KPIs You Need." *Google Cloud*, November 25. https://cloud.google.com/transform/gen-ai-kpis-measuring-ai-success-deep-dive.

37. Grammarly. 2025. "Introducing the Effective Communication Score: A New Way to Improve Business Outcomes." *Grammarly Engineering Blog*, January 22, 2025. https://www.grammarly.com/blog/engineering/effective-communication-score/.

38. Grammarly. n.d. "Introducing Grammarly Tone Detector." *Grammarly Blog*. Accessed Februray 2025. https://www.grammarly.com/blog/grammarly-tone-detector/.

39. The Guardian. 2023. "ChatGPT Shows Gender Bias in Leadership Roles, Study Finds." *The Guardian*. Accessed April, 2024. https://www.theguardian.com/us.

40. Harvard Gazette. "How AI Is Transforming Medicine." *Harvard Gazette*, March 2025. https://news.harvard.edu/gazette/story/2025/03/how-ai-is-transforming-medicine-healthcare/.

41. Harvey. 2024. "Harvey Raises $80M Series C." *Harvey Blog*, August 22. https://www.harvey.ai/blog/harvey-raises-series-c.

42. HCLTech. 2025. *CloudSMART*. HCL Technologies. Accessed April 21, 2025. https://www.hcltech.com/cloudsmart.

43. Hong Kong Institute for Monetary and Financial Research. 2025. *Financial Services in the Era of Generative AI: Facilitating Responsible Adoption*. April 9, 2025. https://www.hkma.gov.hk/eng/news-and-media/press-releases/2025/04/20250409-3/.

44. Huang, Yichi, Shinn Wang, Kevin Lu, Tianjun Zhang, and Pieter Abbeel. 2024. *BARS: A Benchmark for Autonomous Reasoning Skills*. arXiv. June 26, 2024. https://arxiv.org/abs/2406.17910.

45. Huang, Jensen. *NVIDIA Q1 2024 Earnings Call Transcript*. Santa Clara, CA: NVIDIA Corporation, May 2024. Accessed April 2025. https://investor.nvidia.com.

46. HubSpot. 2023. "Working Smarter, Not Harder: HubSpot CRM Introduces New AI-Powered Tools to Boost Productivity and Save Time." *HubSpot Company News*. March 6. https://www.hubspot.com/company-news/hubspot-ai.

47. Hugging Face. *"Hugging Face Model Hub."* Accessed October 2024. https://huggingface.co/models.

48. Hugging Face. *Enterprise Hub Overview*. Accessed April 22, 2025. https://huggingface.co/enterprise.

49. Humlum, Anders, and Emilie Vestergaard. 2024. "Which Workers Are Embracing AI?" *Chicago Booth Review*, November 4, 2024. https://www.chicagobooth.edu/review/which-workers-are-embracing-ai.

50. IBM Institute for Business Value. *Responsible AI: From Principles to Practice*. 2024. https://www.ibm.com/thought-leadership/institute-business-value/report/responsible-ai-2024

51. IDC. 2024. *Worldwide Spending on AI Including Generative AI to Reach $632 Billion in 2028, According to IDC*. May 2, 2024. https://my.idc.com/getdoc.jsp?containerId=prUS52530724.

52. IDC. 2024. "Worldwide Spending on AI Including Generative AI to Reach $632 Billion in 2028, According to IDC." Press release, May 2. https://my.idc.com/getdoc.jsp?containerId=prUS52530724.

53. IDC. 2024. "IDC's Worldwide AI and Generative AI Spending – Industry Outlook." *IDC Blog*. August 21. https://blogs.idc.com/2024/08/21/idcs-worldwide-ai-and-generative-ai-spending-industry-outlook/.

54. IDC. 2024. "China's AI Investment to Reach $26.69 Billion in 2026." *InfotechLead*, December 2024. https://infotechlead.com/artificial-intelligence/chinas-ai-investment-to-reach-26-69-bn-in-2026-idc-82712.

55. IDC. 2025. *Worldwide AI and Generative AI Spending Guide*. Framingham, MA: International Data Corporation. https://my.idc.com/getdoc.jsp?containerId=IDC_P33198.

56. ISO. 2025. "ISO/IEC Guide for AI Compliance: Implementing the EU AI Act." *ISO Standards Documentation.* https://www.iso.org/stand-ard/81230.html.eong, Soyeon, et al. "LLark: A Multimodal Foundation Model for Music." *Papers with Code*, October 11, 2023. https://paperswith-code.com/paper/llark-a-multimodal-foundation-model-for-music.

57. Jagani, Pritesh. 2024. "Last Weekend I Spent 6 Hours Creating This GenAI PM Interview Toolkit." *LinkedIn*, November (approx.). Accessed May 19 2025. https://www.linkedin.com/posts/priteshjagani_last-weekend-i-spent-6-hours-creating-this-activity-7263908955595186176-aPee.

58. Kesler, Aaron. 2025. "Mastering the AI PM Interview: Questions, Strategies, and What Hiring Managers Really Want to See." *LinkedIn*, February 25. Accessed May 19 2025. https://www.linkedin.com/pulse/mastering-ai-pm-in-terview-questions-strategies-what-hiring-kesler-jd1ve.

59. Klee, Miles. 2023. "Google's Bard Chatbot Falsely Claims JWST Took First Exoplanet Picture." *Gizmodo*, February 6. https://gizmodo.com/google-bard-ai-ad-incorrect-webb-telescope-facts-1850087798

60. KPMG International. 2024. "Creating Value with AI Agents: Strategic Priorities for 2025 and Beyond." *KPMG Insights*, September. https://kpmg.com/us/en/articles/2025/creating-value.html.

61. Kurshan, E., J. Chen, V. Storchan, and H. Shen. 2024. "Generative AI from Theory to Practice: A Case Study of Financial Advice." *MIT GenAI*, March 27. https://mit-genai.pubpub.org/pub/l89uu140/release/2.

62. Leviathan, Yaniv, and Yossi Matias. 2018. "Google Duplex: An AI System for Accomplishing Real-World Tasks Over the Phone." *Google Research Blog*, May 8, 2018. https://research.google/blog/google-duplex-an-ai-system-for-accomplishing-real-world-tasks-over-the-phone/.

63. Lewis, Greg. 2025. "LinkedIn Report: How AI Will Redefine Recruiting in 2025." *LinkedIn Talent Blog* (blog). February 13. https://www.linkedin.com/business/talent/blog/talent-acquisition/future-of-recruiting-2025.

64. Li, Nancy. 2024. "What Metrics Would You Define to Measure the Success of a Chatbot You Have Built?" *LinkedIn*, November (approx.). Accessed May 19 2025. https://www.linkedin.com/posts/drnancyli_pmaccelerator-activity-7255219088367534080-eOoK.

65. Liu, Chang. 2025. "Unlocking the Power of Agentic Applications: New Evaluation Metrics for Quality and Safety." *Azure AI Foundry Blog*, April 1, 2025. https://devblogs.microsoft.com/foundry/evaluation-metrics-azure-ai-foundry/.

66. Lundberg, Scott M. 2017. "A Unified Approach to Interpreting Model Predictions." arXiv, May. https://arxiv.org/abs/1705.07874.

67. McKinsey & Company. 2024. *The State of AI in Early 2024: GenAI Adoption Spikes and Starts to Generate Value.* May 21, 2024. https://www.mckin-sey.com/capabilities/quantumblack/our-insights/the-state-of-ai-2024.

68. McKinsey & Company. 2024a. *How Generative AI Could Accelerate Software Product Time to Market.* May 31, 2024. https://www.mckinsey.com/indus-tries/technology-media-and-telecommunications/our-insights/how-genera-tive-ai-could-accelerate-software-product-time-to-market.

69. McKinsey & Company. 2025. "Superagency in the Workplace: Empowering People to Unlock AI's Full Potential at Work." McKinsey Digital, January 27. https://www.mckinsey.com/capabilities/mckinsey-digital/our-insights/superagency-in-the-workplace-empowering-people-to-unlock-ais-full-potential-at-work.

70. McKinsey & Company. 2025. "The State of AI: How Organizations Are Rewiring to Capture Value." QuantumBlack, AI by McKinsey, March 12. https://www.mckinsey.com/capabilities/quantumblack/our-insights/the-state-of-ai.

71. Meisenzahl, Mary. 2024. "Air Canada's Chatbot Gave False Information About Refunds, and a Court Ruled the Airline Is Responsible." *Business Insider*, February 17, 2024. https://www.businessinsider.com/airline-ordered-to-compensate-passenger-misled-by-chatbot-2024-2.

72. Meta AI. *"Llama by Meta."* Accessed October 2024. https://www.llama.com/.

73. Microsoft. 2023a. "Introducing Microsoft 365 Copilot—A Whole New Way to Work." *Microsoft 365 Blog*, March 16. https://www.microsoft.com/en-us/microsoft-365/blog/2023/03/16/introducing-microsoft-365-copilot-a-whole-new-way-to-work/.

74. Microsoft. 2023b. "Providence Uses Azure OpenAI Service to Decrease Clinician Burnout and Expedite Patient Care." *Microsoft Customer Stories*. https://www.microsoft.com/en/customers/story/1701699654934326459-providence-azure-openai-service-united-states.

75. Microsoft. 2024a. "Bringing the Full Power of Copilot to More People and Businesses." *Official Microsoft Blog*. January 15. https://blogs.microsoft.com/blog/2024/01/15/bringing-the-full-power-of-copilot-to-more-people-and-businesses/.

76. Microsoft. 2024b. "Bringing the Full Power of Copilot to More People and Businesses." *Official Microsoft Blog*. January 15. https://blogs.microsoft.com/blog/2024/01/15/bringing-the-full-power-of-copilot-to-more-people-and-businesses/.

77. Microsoft. 2024c. "Physics Wallah: Scaling Personalized Education with Gyan Guru, a Gen AI-Powered Study Companion." *Microsoft Customer Stories*, March 20, 2024. https://www.microsoft.com/en-us/research/blog/microsoft-research-and-physics-wallah-team-up-to-enhance-ai-based-tutoring/.

78. Microsoft. 2024d. "Microsoft and Hugging Face Deepen Generative AI Partnership." *Microsoft Tech Community*. https://techcommunity.microsoft.com/blog/aiplatformblog/microsoft-and-hugging-face-deepen-generative-ai-partnership/4144565.

79. Microsoft. 2024e. "Phi: Small, Open Models Optimized for Reasoning." *Azure AI*. Accessed October 2024. https://azure.microsoft.com/en-us/products/phi.

80. Microsoft. 2025. "Overview of AI in Microsoft Teams for IT Admins." *Microsoft Learn*. Accessed July 6, 2025. https://learn.microsoft.com/en-us/microsoftteams/copilot-ai-agents-overview.

81. Microsoft. 2025. "Microsoft Copilot: Productivity Impact Study—What Copilot's Earliest Users Teach Us about Generative AI at Work." *Work Trend*

Index Special Report. November 15 2023. https://www.microsoft.com/en-us/worklab/work-trend-index/copilots-earliest-users-teach-us-about-generative-ai-at-work.

82. Microsoft and LinkedIn. 2024. "2024 Work Trend Index Annual Report: AI at Work Is Here—Now Comes the Hard Part." *WorkLab.* May 8. https://www.microsoft.com/en-us/worklab/work-trend-index/ai-at-work-is-here-now-comes-the-hard-part.

83. Microsoft Azure. *Microsoft Ignite 2024: Advancing AI Innovation with Azure.* Microsoft. Accessed March 24, 2025. https://news.microsoft.com/ignite-2024/.

84. Microsoft AI. 2023. *Microsoft Responsible AI Dashboard Documentation.* Accessed March 10, 2025. https://www.microsoft.com/en-us/ai/responsible-ai.

85. Microsoft Research. 2025. "AutoGen: Building Multi-Agent Systems with Language Models." GitHub Pages. Accessed February 2025.https://microsoft.github.io/autogen.

86. Mills, Steven, Noah Broestl, and Anne Kleppe. 2025. "You Won't Get GenAI Right If You Get Human Oversight Wrong." *Boston Consulting Group*, March 27, 2025. https://www.bcg.com/publications/2025/wont-get-gen-ai-right-if-human-oversight-wrong

87. Mistral AI. *"Mistral AI: Open-Weight Models for the World."* Accessed October 2024. https://mistral.ai/.

88. MIT Technology Review. 2022. "Meta's Galactica AI Demo Was Shut Down After 3 Days Because It Generated False Information." MIT Technology Review, November 18.https://www.technologyreview.com/2022/11/18/1063515/meta-ai-galactica-shutdown/.

89. Netflix Technology Blog. 2024. "Top Stories Published by Netflix TechBlog in June of 2024." *Netflix TechBlog.* June. https://netflixtechblog.com/archive/2024/06.

90. Nielsen Norman Group. 2024. "Generative UI: Outcome-Oriented Design for AI Tools." *Nielsen Norman Group*, April 21, 2024. https://www.nngroup.com/articles/generative-ui/.

91. Notion. 2023. "Speed, Structure, and Smarts: The Notion AI Way." *Notion Blog*, March 30, 2023. https://www.notion.so/blog/speed-structure-and-smarts-the-notion-ai-way.

92. Notion. 2024. "Notion 2.41: Notion AI Now with Slack, GPT-4, and More!" June 18. https://www.notion.com/releases/2024-06-18.

93. Notion Labs. 2022. "Introducing Notion AI." Notion Blog. November 16. https://www.notion.so/blog/introducing-notion-ai.

94. NVIDIA. "NVIDIA Launches DGX Cloud, Giving Every Enterprise Instant Access to AI Supercomputer from a Browser." *NVIDIA Newsroom*, March 21, 2023. https://nvidianews.nvidia.com/news/nvidia-launches-dgx-cloud-giving-every-enterprise-instant-access-to-ai-supercomputer-from-a-browser.

95. NVIDIA. "CoreWeave Brings NVIDIA H100 GPUs to Developers for AI Inference." *NVIDIA Newsroom*, January 10, 2025.

96. NVIDIA. *Edge AI Solutions for Enterprises*. NVIDIA Corporation. Accessed May 2025. https://www.nvidia.com/en-us/edge-computing/.

97. OECD. 2025. *OECD Regulatory Policy Outlook 2025*. Paris: Organisation for Economic Co-operation and Development. https://www.oecd.org/en/publications/oecd-regulatory-policy-outlook-2025_56b60e39-en.html

98. OpenAI. 2023a. "New Models and Developer Products Announced at Dev Day." OpenAI Blog, November 6. https://openai.com/index/new-models-and-developer-products-announced-at-devday/

99. OpenAI. 2023b. "GPT-4 Turbo and Assistants API: Product Updates and Announcements." *OpenAI Blog*, November 6. Accessed April 22, 2025. https://openai.com/index/new-models-and-developer-products-announced-at-devda

100. OpenAI. 2023c. "GPT-4 System Card." PDF, March 15. https://cdn.openai.com/papers/gpt-4-system-card.pdf.

101. OpenAI. 2024. "Introducing ChatGPT Pro." December 5. https://openai.com/index/introducing-chatgpt-pro/.

102. OpenAI. 2025a. "OpenAI Annual Report 2024: Scaling AI for Global Impact." OpenAI Blog, February 15. Accessed March 2025.

103. OpenAI. 2025b. *Function Calling, Tool Use, and Beyond: Architecting Agents with LLMs*. OpenAI. https://platform.openai.com/docs/guides/function-calling.

104. Ortiz, Sabrina. "What Is ChatGPT Pro? Here's What $200 per Month Gets You." *ZDNET*, December 6, 2024. https://www.zdnet.com/article/openai-introduces-the-new-chatgpt-pro-200-per-month-plan-heres-what-you-get/.

105. PwC. 2023. "PwC Announces Strategic Alliance with Harvey, Positioning PwC's Legal Business Solutions at the Forefront of Legal Generative AI." *PwC Global Newsroom*, March 15. https://www.pwc.com/gx/en/newsroom/press-releases/2023/pwc-announces-strategic-alliance-with-harvey-positioning-pwcs-legal-business-solutions-at-the-forefront-of-legal-generative-ai.html.

106. Rao, Dana. 2023. "Building Safe, Secure and Trustworthy AI: Adobe's Commitments to Our Customers and Community." *Adobe Blog*, September 12. https://blog.adobe.com/en/publish/2023/09/12/adobes-ai-commitments-to-customers-and-community.

107. Reuters. 2025. "South Korea Aims to Secure 10,000 GPUs for National AI Computing Centre." *Reuters*, February 17, 2025. https://www.reuters.com/technology/artificial-intelligence/south-korea-aims-secure-10000-gpus-national-ai-computing-centre-2025-02-17.

108. Reuters. 2025a. "Comment: How Empowering Smallholder Farmers with AI Tools Can Bolster Global Food Security." *Reuters*, January 10, 2025. https://www.reuters.com/sustainability/land-use-biodiversity/comment-how-empowering-smallholder-farmers-with-ai-tools-can-bolster-global-food-2025-01-10/

109. Rooney, Paula. "CIOs to Spend Ambitiously on AI in 2025 and Beyond." *CIO*, May 2, 2025. https://www.cio.com/article/3601606/cios-to-spend-ambitiously-on-ai-in-2025-and-beyond.html.

110. Roza, Nikola. 2025. "Grammarly Statistics, Facts, and Trends for 2025—How Many People Use Grammarly?" Nikola Roza Blog, March 5. https://nikolaroza.com/grammarly-statistics-facts-trends/.

111. Saeedy, Alexander. 2025. "The Rise of Artificial Intelligence at JPMorgan." *The Wall Street Journal*, February 24. https://www.wsj.com/tech/ai/jpmorgan-chase-artificial-intelligence-banking-939b1b32.

112. Salesforce. "Agentforce: The Future of Autonomous CRM." Salesforce, Inc. Accessed April 2025. https://www.salesforce.com/agentforce/.

113. Salesforce. n.d. "Einstein Trust Layer." *Salesforce Help*. Accessed April 30, 2025. https://help.salesforce.com/s/articleView?id=sf.generative_ai_trust_layer.htm.

114. Salesforce. 2023. "Salesforce Announces Einstein GPT, the World's First Generative AI for CRM." *Salesforce Newsroom*. March 7. https://www.salesforce.com/news/press-releases/2023/03/07/einstein-generative-ai/.

115. Salesforce. 2024a. "Salesforce Launches Trusted Generative AI for Customers in Slack." *Salesforce Newsroom*. February 14. https://www.salesforce.com/news/stories/slack-ai-news/.

116. Salesforce. 2024nb. "Salesforce Unveils Agentforce—What AI Was Meant to Be." Salesforce Newsroom. September 12. https://www.salesforce.com/news/press-releases/2024/09/12/agentforce-announcement/.

117. Sequoia Capital. 2024a. "Generative AI's Act Two." *Sequoia Capital*. https://www.sequoiacap.com/article/generative-ai-act-two/.

118. Sequoia Capital. 2024b. "The Copilot Origin Story." *Sequoia Capital Podcast*, August 6, 2024. https://www.sequoiacap.com/podcast/training-data-thomas-dohmke/#the-copilot-origin-story.

119. Spotify. 2023a. "Spotify Debuts a New AI DJ, Right in Your Pocket." *Spotify Newsroom*. February 22. https://newsroom.spotify.com/2023-02-22/spotify-debuts-a-new-ai-dj-right-in-your-pocket/.

120. Shopify. 2023b. "Introducing AI-Generated Product Descriptions Powered by Shopify Magic." *Shopify Blog*. April 20. https://www.shopify.com/blog/ai-product-descriptions.

121. Spotify. 2023c "LLark: A Multimodal Foundation Model for Music." *Spotify Research*. October (26). https://research.atspotify.com/2023/10/llark-a-multimodal-foundation-model-for-music

122. Spotify. 2023d. "Here's What's in Store for Your 2023 Wrapped." *Spotify Newsroom*. November 29. https://newsroom.spotify.com/2023-11-29/wrapped-user-experience-2023/.

123. Spotify. 2025. "Q4 2024 Earnings Report: 675 Million Monthly Active Users." PDF (Shareholder deck), February 7. https://investors.spotify.com/files/doc_financials/2024/q4/Q4-2024-Shareholder-Deck-FINAL.pdf.

124. Siemens. *MindSphere: Powering the Future of Industrial IoT with AI*. Siemens. Accessed April 2025. https://new.siemens.com/global/en/products/software/mindsphere.html.

125. Smith. 2017. "Data Intuition: A Hybrid Approach to Developing Product North Star Metrics." ResearchGate, January. https://www.researchgate.net/publication/322406669_Data_Intuition_A_Hybrid_Approach_to_Developing_Product_North_Star_Metrics.

126. Spehar, Diana. 2025. "AI Governance in 2025: Expert Predictions on Ethics, Tech, and Law." *Forbes*, January 9, 2025. https://www.forbes.com/sites/dianaspehar/2025/01/09/ai-governance-in-2025--expert-predictions-on-ethics-tech-and-law/.

127. Spotify. 2023. "Here's What's in Store for Your 2023 Wrapped." *Spotify Newsroom*. November 29. https://newsroom.spotify.com/2023-11-29/wrapped-user-experience-2023/.

128. Standish, Jill, Brooks Kitchel, and Brett Leary. 2024. *Retail Reinvented: Unleashing the Power of Generative AI*. Accenture. https://www.accenture.com/content/dam/accenture/final/accenture-com/document-2/Accenture-Unleashing-The-Power-Of-Generative-AI-In-Retail-Report.pdf.

129. Stiffler, Lisa. 2024. "Providence Created New Gen AI Tool in 18 Days to Speed and Improve Responses to Patient Messages." *GeekWire*, February 20, 2024. https://www.geekwire.com/2024/providence-created-new-gen-ai-tool-in-18-days-to-speed-and-improve-responses-to-patient-messages/.

130. TechTarget. n.d. "Multimodal AI and the Rise of Orchestration Platforms." Accessed March 2025. https://www.techtarget.com/.

131. Turovsky, Barak. 2024. "How to Evaluate Generative AI Use Cases." *Creator Economy*, April 2024. https://creatoreconomy.so/p/how-to-evaluate-generative-ai-use-cases.

132. Verma, Rohit. 2024. "As the Product Manager at Duolingo, How Would You Leverage Generative AI to Enhance the User Experience?" *Bootcamp* (blog on Medium), November 18. Accessed May 19 2025. https://medium.com/design-bootcamp/pm-interview-question-as-the-product-manager-at-duolingo-how-would-you-leverage-generative-ai-to-e8f56122340f.

133. Vincent, James. 2023. "Samsung Bans ChatGPT After Employee Leaks Code." *The Verge*, May 2. https://www.theverge.com/2023/5/2/23707796/samsung-ban-chatgpt-generative-ai-bing-bard-employees-security-concerns

134. Volberda, Henk. 2025. "AI Will Perform Most Work Tasks by 2030." *Amsterdam Centre for Business Innovation*, January 8. https://acbi.uva.nl/content/news/2025/01/ai-will-perform-most-work-tasks-by-2030.html.

135. Wall Street Journal. 2025. "Venture Market Notches Strong Quarter—Thanks Mostly to One Mammoth AI Deal." *The Wall Street Journal*, May 2025. https://www.wsj.com/articles/venture-market-notches-strong-quarterthanks-mostly-to-one-mammoth-ai-deal-668f056c.

136. Wiggers, Kyle. 2023. "Hugging Face Raises $235M from Investors Including Salesforce and Nvidia." *TechCrunch*, August 24. https://techcrunch.com/2023/08/24/hugging-face-raises-235m-from-investors-including-salesforce-and-nvidia/.

137. Wiggers, Kyle. 2025. "Anthropic appears to be using Brave to power web search for its Claude chatbot." *TechCrunch*, March 21, 2025.

https://techcrunch.com/2025/03/21/anthropic-appears-to-be-using-brave-to-power-web-searches-for-its-claude-chatbot/

138. Wilner, Michael. 2025. "Order Imposing Sanctions for Submission of Brief with AI-Generated Fake Citations." *United States District Court for the Central District of California*, April 15, 2025. https://www.theverge.com/news/666443/judge-slams-lawyers-ai-bogus-research

139. Workday. *Workday Unveils Next Generation of Illuminate Agents to Transform HR and Finance Operations*. Workday Newsroom, May 19, 2025. https://investor.workday.com/2025-05-19-Workday-Unveils-Next-Generation-of-Illuminate-Agents-to-Transform-HR-and-Finance-Operations.

140. World Economic Forum. 2024. "How We Can Best Empower the Future of Business in APAC." *World Economic Forum*, June 2024. https://www.weforum.org/stories/2024/06/how-we-can-best-empower-the-future-of-business-in-apac.

141. World Economic Forum. 2025a. "Why AI Infrastructure and Governance Must Evolve Together." *World Economic Forum*, May 2025. https://www.weforum.org/stories/2025/05/why-ai-infrastructure-and-governance-must-evolve-together.

142. World Economic Forum. 2025b. *The Future of Jobs Report 2025*. Geneva: World Economic Forum. https://www.weforum.org/publications/the-future-of-jobs-report-2025.

143. World Economic Forum. 2025c. *State of Global AI Regulation: Trends, Risks, and Opportunities*. Geneva: World Economic Forum. https://www.weforum.org/reports/state-of-global-ai-regulation-2025

144. World Economic Forum. 2025c. *AI in Action: Beyond Experimentation to Transform Industry*. Geneva: World Economic Forum. https://reports.weforum.org/docs/WEF_AI_in_Action_Beyond_Experimentation_to_Transform_Industry_2025.pdf

145. xAI. 2025. "Grok 3: Multimodal AI for Enterprises." Accessed March 2025. https://x.ai/grok.

146. Zapier. "Connect Zoom to Anthropic (Claude) with AI." Accessed April 22, 2025. https://zapier.com/apps/zoom/integrations/anthropic-claude.

147. Zhang, Kang, Xin Yang, and Shengyong Yang. "Artificial Intelligence in Drug Development." *Nature Medicine* 31, no. 1 (January 2025): 1–13. https://www.nature.com/articles/s41591-024-03434-4.

Acknowledgements

Writing this book was never a solo effort. It was shaped by the insights, encouragement, and support of many people who influenced not only the words on these pages but also the journey behind them.

To the product managers, engineers, researchers, and peers who shared their stories and experiences—thank you for grounding this book in the realities of building AI products. Your perspectives made it sharper, more honest, and more useful.

To the industry leaders who contributed their reflections—Amit Ghorawat, Shambhavi Rao, Selena Zhang, Sunny Tahilramani, Rocky Zhang, Jeremy Schifeling, Mark Cramer, and Saty Das—your wisdom added real-world depth and nuance to the lessons in this book.

To the readers and practitioners who shared their personal transition journeys into GenAI product roles—thank you for trusting me with your stories. Your courage, clarity, and openness reminded me why this book needed to be written.

To the endorsers who believed in this work early—Lewis C. Lin, Matt Schnugg, Vamsi Krishna, Suvrat Joshi, Abhishek Sharma, Umang Goyal, Partho Sarkar, Nikhil Sharma, and others—your support gave this book the credibility and momentum it needed to reach more people.

To my readers-turned-friends, aspiring PMs, and career switchers who reached out with questions, doubts, and dreams—this book exists because of you. You reminded me how important it is to create resources that empower.

To my colleagues at Microsoft—thank you for the support, collaboration, and inspiration. Working alongside people so passionate about building the future of AI made this journey richer.

To the early reviewers, mentors, and editors who helped shape rough drafts into sharper ideas—your feedback made this book better at every turn.

And to everyone quietly cheering from the sidelines—you know who you are. Your belief helped bring this vision to life. Thank you for being part of this journey.

About the Author

Saumil Shrivastava is a product leader at Microsoft, where he drives AI product strategy and ecosystem development on the Azure AI platform team. He works at the intersection of cutting-edge research, large language models, and developer tools—helping startups and enterprises build, deploy, and scale GenAI solutions that solve real user problems and deliver tangible value. He also leads Foundry Labs, the front door to Microsoft Research for developers, startups, and enterprises.

With over a decade of experience launching 0-to-1 products across startups and Big Tech, Saumil has led cross-functional AI initiatives and coached dozens of product managers transitioning into AI. Before rejoining Microsoft's AI organization, he held senior product leadership roles at UiPath and helped scale Power Platform's maker and admin experiences during his earlier years at Microsoft.

A passionate mentor, Saumil has guided over 500 aspiring and practicing product managers around the world—offering structured support on career transitions, interviews, and breaking into AI product roles. (You can connect with him at calendly.com/pmsaumil.)

Saumil earned his MBA from the Ross School of Business, University of Michigan, and his undergraduate degree from IIT Bombay, one of India's most prestigious engineering institutes. He is also the author of *A Roller Coaster Ride*, a national bestselling novel that reached thousands of readers across India.

Beyond product leadership, Saumil shares curated content on well-being and life transformation through his Instagram channel @GoodVibesBySam, which has inspired thousands seeking a more intentional, positive life. You can also follow his writing and reflections on Linkedin (@saumilshri), Medium (@saumil23) and Twitter (@sam1983).

Zero to GenAI Product Leader is his most practical and personal book yet—born from real-world experience, community conversations, and the belief that anyone can build a future in AI with the right mindset and map.